The Alchemy of Us

The Alchemy of Us

How Humans and Matter Transformed One Another

Ainissa Ramirez

The MIT Press

Cambridge, Massachusetts | London, England

MIT Press first paperback edition, 2021
© 2020 Ainissa Ramirez

This book was set in ITC Stone Serif Std and ITC Stone Sans Std by Toppan Best-set Premedia Limited. Printed and bound in the United States of America.

Library of Congress Cataloging-in-Publication Data

Names: Ramirez, Ainissa, 1969– author.
Title: The alchemy of us : how humans and matter transformed one another / Ainissa Ramirez.
Description: Cambridge, Massachusetts : The MIT Press, 2020. | Includes bibliographical references and index.
Identifiers: LCCN 2019029157 | ISBN 9780262043809 (hardcover) ISBN 9780262542265 (paperback)
Subjects: LCSH: Materials—History—Popular works. | Inventions—History—Popular works. | Technology—Social aspects—Popular works.
Classification: LCC TA403.2 .R36 2020 | DDC 620/.11—dc23
LC record available at https://lccn.loc.gov/2019029157

10 9 8 7 6 5 4

For my mother and grandmother.

All that you touch
You Change.

All that you Change
Changes you …

Octavia E. Butler

Contents

Introduction

Ever since I was four I wanted to be a scientist, which made me an unusual little girl in my corner of New Jersey. I was an inquisitive youngster who wanted to know why the sky was blue, why leaves changed color, and why snowflakes had six sides. With this perpetual curiosity, I eventually got the idea of becoming a scientist from watching television shows during the late seventies and eighties. Back then, I loved programs like *Star Trek* (with Spock), *The Bionic Woman*, and *The Six Million Dollar Man*, but the show that solidified my path toward science was a public television show called *3-2-1 Contact*. One of the repeating segments had a young African American girl solving problems, and when I saw her using her brain, I saw my reflection.

Science in my childhood was full of fun and wonder. Years later, however, my dreams of becoming a scientist nearly flatlined, as I sat in a science lecture hall with tears welling up in my eyes. The science lectures were far from being any fun or bringing any wonder. In fact, these classes were dry and the lessons were designed to weed out students. Chemistry courses were prescribed cookbooks; engineering exercises examined the steam engine; and math instruction was unmotivated. I knew these subjects were better than how they were being taught, and I suffered through with the help of mentors, tutors, and many hours in the library. Fortunately, I found a major that returned my wonder—a little-known field called materials science, where I learned that everything in our world is the workings of atoms.

Materials science is a bit like my home state of New Jersey, because it is wedged between two more well-known entities. For New Jersey, they are New York City and Philadelphia. For materials science, they are chemistry and physics, and just like New Jersey, materials science has not been able to make a case for itself of how great it is on its own. If there were no City of Brotherly Love or Big Apple, New Jersey would have been a fine and respectable state. It might have done well if it was located somewhere west, like next to Iowa, for New Jersey has its own history, its own culture, and certainly its own attitude. But the Garden State is overshadowed by its overpowering neighbors. The same holds for materials science.

Despite my penchant for unappreciated states and science majors, I loved materials science, in part because my college professor at Brown said something that blew me away. "The reason why we don't fall through the floor, the reason why my sweater is blue, and the reason why the lights work is because of the way that atoms interact with each other," said Professor L. Ben Freund. "And if you can find out how they do that, you can also change the way that atoms act to make them do new things." After he said that, I looked at everything around me in a new light. I stared at my pencil, which was able to make a mark because layers of carbon atoms slid over each other. I looked at my glasses, which helped my wretched eyes see, because the glass bent light to my distant retinas. I looked down at the rubber in my shoes, which brought a springy comfort to my feet, because of the twisted and coiled molecules inside them. Atoms did all this. This guy was telling me something that made the whole world make sense to me. My wonder was back, but it took most of my undergraduate years to get in touch with it again.

Had I been weeded out by those hard, introductory science courses, this opportunity could easily have been lost. When I graduated, I swore that I would do whatever I could to make sure that no one else suffered through science that way. This book is my attempt to fulfill that old promise.

Two decades later, long after I became a scientist, the idea for this book unexpectedly came to me. Here, as an adult, I was still fond of wonder, but I preferred the type that was part learning and part thrilling. Glassblowing fit that definition, so I signed up to take a few classes.

I was filled with awe in my glassblowing class, like when I watched my instructor, Ray, transform a clear lump into a galloping horse with a few short tugs. But I was also full of fear, like when Ray warned me that glass drips on the floor could melt a hole in the bottom of my shoe. Working with the glass, I gained a deeper appreciation for it than I had acquired in my studies. But soon I would learn something that I didn't expect.

One Wednesday night I arrived at my evening glassblowing class in a full-blown work funk. Usually, when I went to glass class, I treated the molten liquid with the utmost respect. As in every other class, I would dip my pipe into the vat, pull out a small dollop, blow a small golf-ball sized bubble, and shape a small vase. But during class that night, I didn't care to be so safe.

On that New England winter evening, I grabbed triple the amount of glass on the end of my pipe dripping some on the floor, and nearly maxing out my muscles. I didn't care. I puffed a bubble the length of two baseballs, and heated and shaped and heated and swung and heated and shaped. When my funk finally lifted, I noticed that this vase was one of the best specimens I had ever made. As I neared the final stages, I held my pipe with the vase inside the furnace, and struck up a conversation with a classmate. I began to think about other things besides my glass piece—and that's a big no-no.

As I chitchatted, the vase stayed in the furnace too long and came out an incandescent orange, hunching down at the end of my pipe. My pride for it evaporated. I spun the glass 180 degrees. But the vase countered by hunching down on its new lower side. I turned again. It hunched. I turned again. It hunched. I turned again. It hunched again. Sweat began to bead on the top of my lip.

My hope was that a winter breeze from the open window would save me from my dilemma by cooling and hardening the glass vase. But the studio's furnaces kept the inside climate tropical. I was in trouble, and the glass seemed to sense it.

Eventually, the vase took matters into its own hands. When I spun the pipe again, the vase dove to the ground. I checked my skin for glass shards and was fine. But the vase, throbbing on the cluttered floor, wasn't.

I yelled to my instructor, Ray, who swept in wearing asbestos gloves. He scooped up the vase, reattached it to my pipe, put it in the furnace, and brought it back to my bench. Rolling the pipe back and forth, Ray carved open the vase's sealed lips, and rounded the flattened side with a wet wood block. The glass was going to survive, but with a new physique.

After both the glass and I cooled off, I had a moment to contemplate what had happened and an idea came to me. I shaped the glass, and the glass was shaping me. I was giving it form, even when I dropped it, and the making of a glass vase on this Wednesday night was not only distracting me from a bad day, but forging a deeper understanding and appreciation of glass, and materials in general. Perhaps this thinking was a bit existential, but the event inspired this book. The notion that materials and humans are being molded by each other catalyzed me to explore how materials in history shaped us.

The Alchemy of Us shows how materials were shaped by inventors, but also how those materials shaped culture. Each chapter is titled with a verb to demonstrate how the meaning of that word was fashioned. Particularly, this book highlights how quartz clocks, steel rails, copper communication cables, silver photographic films, carbon light bulb filaments, magnetic hard disks, glass labware, and silicon chips radically altered how we interact, connect, convey, capture, see, share, discover, and think. *The Alchemy of Us* fills in the gaps of most books about technology by telling the tales of little-known inventors, or by taking a different angle to well-known ones. I chose to look at the gaps, at the silences in history, because they too are instructive about the makings of our culture. I highlight "others" to allow more people to see their reflection. I use storytelling with the hopes of bringing the wonder and fun of science to more people.

My wish is that you will come away with an appreciation of the technologies that abound—and also inherit a sense of urgency. In order to create the finest versions of ourselves, we need to think critically about the tools that surround us. This book aims to nurture that perspective. In these pages, you will pick up plenty of talking points for party conversations, but also some cud to chew.

Overall, *The Alchemy of Us* seeks to create a new connection to the world, to history, and to each other. Admittedly, the linkage between science and culture may seem a heady concept, but an erudite twentieth-century sociologist named Madonna sang about it when she crooned that we live in a material world. She was absolutely right. Everything around us is made of something. But not only do we live in a material world, we are in a dance with these materials, too. We form them, but they, in turn, shape us. This was the lesson that my deformed vase was trying to impress upon me on that wintery Wednesday night. Let's benefit from its sacrifice and see how.

Ainissa Ramirez, PhD
New Haven, CT

1

Interact

How better clocks, made possible by small metal springs and vibrating gems, helped us keep time, but also made us lose track of something precious.

Ruth and Arnold

Like clockwork, a recognizable knock came at the door. It was a Monday in the fall of 1908 and just like every Monday, a woman named Ruth Belville stood at the entryway of a London watchmaker. She wore a dark dress cinched with a broad waistband that offered a hint there was a slim shape underneath the thick fabric. Her ankle-length hem cast a wide shadow that obscured her shoes from sight. Her hair was gathered up neatly under her hat and on her arm hung a modest yet oversized handbag. There at the entrance, aware of the time, she eagerly waited. When the door finally opened, the store-keeper greeted the weekly visitor with "Good morning, Miss Belville. How is Arnold today?" She replied, "Good morning! Arnold is four seconds fast." She then reached into her handbag, grabbed a pocket watch, and passed it to the watchmaker. He used it to check the store's main clock and then returned the pocket watch to her. She left. Their transaction was now complete. Ruth Belville was in the unusual business of selling time with her watch named Arnold.

In the early twentieth century, the world struggled to know what time it was. The early sundial and water clock, and later the hourglass, showed the marching of time, with the motion of a shadow, the lowering of a liquid's surface, or the filling of space by sand. But knowing the exact hour and minute of the day required astronomical observations and calculations. Such information lived in observatories, such as the Royal Observatory in Greenwich, England. To know the exact time of the day, visitors would have to make a trip to Greenwich to this astronomical haven to obtain it.

Numerous businesses required knowing the precise time. As one would expect, railroad stations, banks, and newspapers needed to know what time it was. But they were not alone. Taverns, bars, and pubs needed to know the time, too, since strict laws passed in the 1870s in England prevented the sale of alcohol after set hours. Those who did not comply risked losing their license and livelihood. All these disparate businesses in London required the observatory's precise time, but they did not have the luxury of making the several-mile journey to get it.

Ruth Belville (1854–1943) brought time to her clients. Once a week, she would take the three-hour journey from her cottage in Maidenhead, thirty miles west of London, and then travel to Greenwich to the Royal Observatory. Reaching its gate by nine o'clock, she rang and was greeted by the gate porter, who formally invited her in. An attendant approached her and she handed over her watch, Arnold. While she waited and had a spot of tea and chitchatted with the gate porter, her watch was compared to the observatory's master clock. The attendant then returned Arnold to her along with a certificate stating the difference between its time and their main clock. With her trusty timekeeper and an official document in hand, Ruth made her way down the hill, walked over to the Thames to catch a ferry, and proceeded to her customers in London.

Ruth Belville brought time to her customers as society's habit of living by the clock came into full bloom. Life before clocks was different, and this evolution can be likened to what we experience as we mature from children to adults. When babies are born, they have their own clock—mealtime, sleeptime, playtime. But as we mature, life divorces itself from those biological cues as we adhere to the clock with school starts, recess, and dismissals. Society went through a similar metamorphosis, switching from nature's cues to the clock's. Originally, the sun was the main means of timekeeping—with sunrise, high noon, and sunset. Before the clock, society didn't have appointments within those bounds. The clock allowed us to meet and interact with each other at any time, but brought with it what Aldous Huxley called the "vice of speed." Before clocks, we would wait for long periods for someone to arrive. Today, in the United States, we will no longer wait more than twenty minutes after a designated appointment time. Accurate timekeeping changed society and touched all aspects of life. One of these changes brought by timekeeping keeps us up at night. Living by the clock has changed how we sleep.

A Distant Sleep

Our ancestors slept differently. They didn't sleep longer. They didn't sleep better. What they *did* do was sleep in a manner that wouldn't really be recognizable to us today. Before the Industrial Revolution, our ancestors slept at night in two separate intervals. If we looked back, we would see them retiring for the evening around 9 or 10 p.m. and sleep for three and a half hours. Then, unceremoniously, they'd wake up after midnight and stay up for about an hour or so. When they grew tired again, they would return to bed and doze off for another three and a half. These distinct doses of slumber were known as "first sleep" and "second sleep," and this was the customary way of catching z's.

Unlike how we feel about sleep today, our ancestors were neither anxious about waking up in the middle of the night, nor worried about having a medical condition. In fact, they felt the opposite about being awake: they savored it. They used their sleep half-time to write, read, sew, pray, pee, eat, clean, or gossip with next-door neighbors (who, chances are, were up in the wee hours before dawn, too). As soon as this midnight cohort felt drowsy again, the intermission was over, and they made their way back to bed to continue sleeping for the second act.

While sleeping in intervals, or segmented sleep, seems surprising to us today, it is actually quite old—over two thousand years old, at least. Since very few people remember segmented sleep, the best evidence for it resides in early books. Ancient texts like Homer's *Odyssey* (written around 750 B.C.) and Virgil's *Aeneid* (19 B.C.) mention "first sleep." "First sleep" is also mentioned in a number of classics, such as *Don Quixote* (written in 1605), *The Last of the Mohicans* (1826), *Jane Eyre* (1847), *War and Peace* (1865), and Charles Dickens's *The Pickwick Papers* (1836). Over a thousand newspapers from the nineteenth century cite first sleep and second sleep hundreds of times, too.

Segmented sleep was part of everyday life in Western culture. But by the early twentieth century, it was gone. The Industrial Revolution changed our sleeping patterns with a one-two punch: the first strike was tangible and direct, with the invention of artificial lights; the second blow was cultural and subtle, with the desire for punctuality brought by the clock. When artificial lights came into being, they pushed back the darkness and lengthened the day. Additionally, we grew obsessed with time, with being on time, and with not wasting time. As such, it was only a matter of time before this compulsion effected how we sleep.

When the Puritans arrived in North America in the seventeenth century, they brought many things with them. One of those was their time-sense and the belief in using time wisely. Later, these religious values were transmogrified with capitalism to the Benjamin Franklin adage that "time is money." With this thinking, we as a culture grew increasingly time conscious and our actions and interactions were ordered by it. At the heart of our culture was the factory, and the clock gave it its pulse. Clock bells rang out telling workers when to start, when to stop, and even when to produce faster. This pulse didn't beat solely within the factory's walls, though. Family life began to center around the factory, too. All events within the home accompanied that pulse, such as when to wake, when to eat, when to leave, when to return home, and when to turn in for the night.

For people born and bred in this modern day, our burgeoning obsession with time in the nineteenth century may be difficult to imagine. Ruth Belville's time-distribution business was of her era. One way to illustrate our newfound time mania rests in the creation of new words. For example, in sports, we paused a game with half-time (in football in 1867) or a time-out (1896) in other sports. Popular books in science fiction, like H. G. Wells's *The Time Machine* (1895), got us excited about time travel. Countries formed a worldwide network of synchronous clocks using Greenwich Mean Time (established in 1847) with the creation of standard time (1883), so we gained timelines (1876), time zones (1885), and time stamps (1888). People became aware of their mortality and described things by their time span (1897) or their time limit (1880). We were cognizant if something was old-fashioned and noted it to be behind the times (1831). When someone was sent away to prison they were said to be doing time (1865). Mostly, though, we lived in a society that was timewise (1898), kept to a timetable (1838), and desired to make good time (1838). We as a society increased our awareness of time. All aspects of our lives were touched by it, including our sleep.

When Ruth Belville got into the unusual business of selling time, her customers, sleeping differently than today, had a growing desire to know what time it was. For her line of work, Ruth was called the Greenwich Time Lady, bringing her watch with accurate time to those needing to know it. Ruth was able to offer her services with her watch, Arnold, but Ruth was not Arnold's first carrier. Her mother did the same work until her death and her father started this peculiar business before his widow. In total, Ruth's family was in the business of purveying time for nearly 104 years.

The Belvilles fell into this line of work accidentally. Ruth's father, John Henry Belville, was an affable man who acquiesced to the mounting work at the observatory as a meteorologist and astronomer. The demanding leadership grew increasingly frustrated by the numerous interruptions caused by local astronomers desperate to know the precise time for their observational work. Instead of having unannounced visitors coming to the observatory and disrupting their scientific activities, the plan was to bring time to those who desired it. The dedicated and gentle-mannered John Belville provided time to his nearly two hundred customers.

On July 13, 1856, John Belville died and left his watch to his third wife, Maria Elizabeth. She needed to find a way to support herself and her two-year-old daughter, Ruth, since her husband left no pension. So she sold time for the rest of her life to her one hundred customers. Arnold the watch was then passed down in 1892 to Ruth, who at age thirty-eight continued the family business.

Arnold was formally known as John Arnold No. 485 and named after its maker, who built it in 1786. This was a highly accurate chronometer, better made than a standard pocket watch. Legend has it that Arnold was originally designed as a gift for royalty, specifically for the duke of Sussex, a son of George III. The duke of Sussex thought the watch was too big, though; he likened it to a "warming pan" and refused it. As luck would have it, the duke of Sussex was connected to the Royal Observatory, which put Arnold on a path toward John Belville's hands when the time-distribution service was created. Arnold was originally in a gold case, but Ruth's father, John Belville, had it rebuilt and updated in a silver case so it would be less attractive to thieves. Arnold's beauty wasn't on the outside, however, but on the inside. Behind Arnold's white enamel face and gold hands sat a range of materials working in sync—brass gears, ruby pivots, and a steel spring. This timepiece from the eighteenth century clicks five times a second, which is a masterful performance even today.

Arnold the watch is part of an old tradition, for the telling of time has been a human pursuit from the beginning of antiquity. Sundials and water clocks give a sense of the hours passing. But to measure time like a yardstick, what was needed was a regular pattern that could be counted. According to legend, Galileo observed that the lamps in a cathedral in Pisa swung regularly. Using his pulse, he found that they moved back and forth with a stable and unchanged rhythm—a natural frequency. This simple observation was what society had been waiting for as a way to measure time. Soon, pendulum clocks and later smaller pocket watches, like Arnold, would use springs inside to provide the clock's ticks. But making small watches like Arnold would not be easy, for the spring inside had to be uniformly made to tell time accurately. Making watches was incredibly frustrating. One eighteenth-century clockmaker in England became so annoyed with his clocks that he did something about it.

Benjamin Huntsman's Clocks

Benjamin Huntsman was irritated by his timepieces. Born in 1704 in Epworth, England, Huntsman was known to be a clever, inventive, and handy clockmaker. In his village, he was a fixer of all things mechanical, from locks to clocks to tools to rotisseries. Yet, despite all his technical ability and acumen, he frowned on his clocks. They were poor timekeepers because their metal springs were inferior.

Deep in a clock is a gadget that ticks. In some clocks, a pendulum swings back and forth; in small pocket watches, it is a combination of a spiral metal spring and a balance wheel. The spring is coiled so that it expands and contracts, like a chest cavity, producing the clock's ticks and tocks. Clocks with springs that hiccup give a fast time; clocks with springs that breathe deeply supply a slow time. Accurate clocks require metal springs that are flexible and flawless, inhaling and exhaling steadily.

Unfortunately, the quality of the metals at hand for Huntsman was inconsistent, because the ingredients were not evenly mixed. Plus, these metals contained unwanted particles. The poor mixing of ingredients made watches behave irregularly and the particles caused the springs to snap. Neither boded well for good timekeeping.

In his quest to make better springs for his watches, Huntsman turned his attention to their starting material, a metal called blister steel. Blister steel was made by adding carbon to iron, which steelmakers accomplished by placing iron bars into a furnace, heating them until they were red-hot, and then surrounding them with bits of charcoal. After five days, the bars contained a lot of carbon from the charcoal, with most of it near the surface as on a poorly marinated steak. To mix these ingredients further, steelmakers had to heat the metal to soften it, beat the metal with a hammer to flatten it, and then fold the metal onto itself to blend it. This method was certainly beneficial for incorporating carbon, but it did nothing for getting rid of the unwanted particles. Huntsman had to come up with another way.

One day in his clockworks in Doncaster, Huntsman had a simple, but revolutionary, idea: melt the metal fully. When the metal was molten the ingredients would intermingle better and deliver a uniform mixture of carbon. Plus, the unwanted particles, which were lighter than the molten liquid, would float to the top, separating like oil and water, which Huntsman could then remove.

Beginning in secret and working with little interaction with the outside world, Huntsman failed hundreds of times. Although the records of his research were burned in a fire, artifacts of his attempts were buried outside his works in Doncaster. In these unsuccessful experiments, he aimed to blend in the carbon and he strived to eliminate those unwanted particles. After ten years of work, around 1740, Huntsman finally perfected his steel. He commemorated his achievement by making a clock.

The secret behind Huntsman's success was the creation of a container to carry the molten metal. This container, or crucible, looked like a tall, ancient vase and was made of a ceramic that could withstand the fiery metal's heat as well as sustain the heavy metal's weight. To make his crucible he ground up imported pots from Holland and added graphite and a special clay from England called Stourbridge clay. Next, he added water to the blend and then one of Huntsman's trusted workmen stomped on the mixture with his bare feet for eight to ten hours. Bare feet permitted both the squeezing out of air pockets and the detecting of pebbles in the clay, either of which would cause crucibles to crack and leak out the steel.

After the clay was kneaded and shaped, the resulting containers were dried and then fired in a kiln. When these ceramic crucibles were completed, the steelmaking began.

Huntsman perfected his method in his new works near the city of Sheffield (a steelmaking hub). His workmen placed small bits of blister steel into a crucible and then lowered the crucible into a furnace, where it remained for five hours. When the crucible came out, a workman skillfully poured the steel into a mold, making sure to prevent the unwanted floating layer from falling in, too. When the pouring was complete, the final metal in the mold was crucible steel—a uniformly blended metal that was fashioned into fine watch springs that expanded and contracted consistently. Benjamin Huntsman's invention made for better watches, ones that could be carried in a pocket or hung on a wall or carried throughout London to provide the time, like Ruth Belville's Arnold.

After Ruth Belville acquired the accurate time from the Royal Observatory, she and Arnold headed toward the London Docks, and then crisscrossed the city and cut through different parts of society along the way. She started in the east to provide time to the vice-ridden and odorous docks. Tick. After that, she shot over to the fashionable west end of Oxford, Regent, and Bond Streets, to posh shops and fine jewelers (including the royal jeweler). Tick. Next, she turned north to Baker Street, to factories and commercial buildings. Tick. Then, she headed south, to individual customers in the suburbs. Tick. Later, she brought time to two millionaires who relished the status symbol of GMT in their homes. Tick. All the while, she intersected the center of London to provide time to banks. Tick. At last, she ended that long day and returned home to Maidenhead only to repeat the cycle again in seven days. Tock.

With Arnold in her handbag, she walked and walked over cobbled-stone roads coated with coal dust and dotted with horse manure. When she could afford it, she would take public transportation, including the tram, subway, and train. Life in the city was hard, dirty, and teetering at a flash point. The air was thick with smoke and fog. And there was a constant cacophony of shouting vendors, the click-clack of horses, and the rumble of an occasional motorcar as she carried the time. Ruth walked through miles and miles of the city eking out a living. She was a businesswoman in a time when women could not vote. Described as hale and hearty, Ruth was emotionally tough and resolute, and she had a common touch as she moved through a world dominated by men. Together, Ruth and Arnold were trusted fixtures in London life.

Near the end of Belville's career another clock was installed in the United States toward which people started to migrate. Ruth was a one-woman operation who made her way up a steep hill to the outskirts of the city to get the precise time. Toward the end of her career, crowds of New Yorkers headed downtown to get the precise time, too.

In 1939, at 195 Broadway, at the corner of Fulton Street in Manhattan, an art deco installation was placed in the window of the AT&T corporate headquarters. It was a clock, but not just any clock. It was touted as the most accurate public clock in the world. Every day, particularly between noon and 2:00 p.m., hundreds of pedestrians made their pilgrimage to the window, stopped, and held their finger on their watch stem waiting for the sweeping second hand to reach the top, so they could set their watch accurately. Unbeknownst to these time-seekers was that a little-known scientist had made this clock possible by superseding Benjamin Huntsman's springs. The ticking behind the nearly three-foot clock face was provided by a piece of quartz that had a special behavior to it. The wrangler of this quartz gem was scientist Warren Marrison.

Gems That Jiggle

Warren Marrison was a clever and quiet Canadian boy who, in the 1910s, seemed out of place and of a different time. He would spend his life trying to correct this. He was born in 1896 in Inverary, Ontario, Canada, and his first major accomplishment was escaping his father's bee farm, for the younger Marrison possessed aspirations greater than being an apiarist. In this small horse-and-buggy town, Warren dreamed of electricity and worked hard in school to get to America to fulfill his vision for the future. He would get his wish.

By 1921, he and his new wife moved to New York City so Warren could start his career at Western Electric's Engineering Department (later renamed Bell Laboratories). Bell Laboratories was the research arm of the telephone company. American Telephone and Telegraph (or AT&T) absorbed Bell Labs from Western Electric a few years after Marrison arrived. Its building was located at 463 West Street at the corner of Bethune Street—a thirteen-story construction that still stands today. The elevated railroad, the structure of which remains as the High Line, pierced through its third floor a decade after Marrison arrived, bringing a periodic tremor to the building. This Bell Labs edifice was a concrete skyscraper that lacked great beauty but, fortunately, imagination thrived inside.

Within the building was a constant buzz of activity, with scientist "worker bees," like Marrison, in suits and ties. The floors were hardwood maple, the walls were bare, and the windows were numerous, offering plenty of light to avoid the expense of electric lights. Warren Marrison worked on the seventh floor, where his lab benches overflowed with specialized tools and scientific instruments that exposed their guts of wires and electronics. Scientists were required to put in a five-and-a-half-day workweek, but research never kept to the clock.

As the Marrison family grew, it moved to a home in Maplewood, New Jersey, in the late 1920s. Marrison would pour honey into his tea and tell his two daughters about the buzzing bees of his youth on the farm. Marrison smiled easily and despite his relatively small stature, at 5' 10", he had a booming voice. Often he was unaware of its potency and embarrassed his youngest daughter when he explained science to her in the library. His excitement for science, like his voice, was hard to tame.

At Bell Labs, Marrison's projects zigzagged between many different inventions: He worked on adding sound to movies. He developed ways of sending moving pictures across airwaves to create television.

Day and night, he endlessly filled his lab notebooks with ideas that contained sophisticated circuit diagrams that allowed electrical signals to "talk" to mechanical parts. Soon, Marrison became enamored with using quartz for clocks.

Bell Labs had one of the first radio stations: WEAF, and the idea for the quartz crystal clock actually came from radio. Radio stations broadcast at specific frequencies designated by the numbers seen on a radio dial. It was hard, however, for radio stations to know if they were broadcasting at the right frequency, which was necessary to avoid interfering with neighboring stations. Marrison's project in 1924 focused on making a machine that produced accurate and unwavering signals that served as a standard frequency. Using a large quartz gem, Marrison sawed a sliver off and then mounted it into his electronics. Quartz is an unassuming mineral with an extraordinary secret—it vibrates when zapped by an electrical circuit. The quartz slice gave a beat at a specific rate, which became the radio station's standard. Marrison's frequency generator served as a North Star for those lost in a sea of radio waves.

After his success with the radio standard, Marrison had another idea. Instead of using the vibrating crystal to send a precise radio signal, he would vibrate the crystal, which produced a known number of vibrations in a second, and count the vibrations to mark off time. This way the counts become the "measuring stick" of time. With that idea, Marrison convinced natural quartz to beat. By fashioning a crystal into a doughnut shape, he was able to make quartz flap up and down like a drum head. His quartz rang one hundred thousand times in one second and those vibrations were counted to tell time. Quartz was able to do this because of its little known secret. It dances with electricity because of a weird phenomenon called *piezoelectricity*.

Piezoelectricity was discovered by Pierre and Jacques Curie in 1880 in Paris. While in their early twenties, these young men desired to make a name for themselves in the very crowded field of mineralogy. At the time, many other scientists excavated gems from the earth and studied and classified their color, clarity, and faceting. The Curie brothers wanted to go further and see what else these gems could do in different situations. Pierre was always fascinated with the symmetry of geometric shapes, particularly in minerals. Quartz lacks the simple symmetry of other gems, like diamond or salt. Facets on one side of a quartz gem were not matched with an identical facet on the other side. This meant that the atoms inside didn't have a mirror image, and that physical properties usually cancelled out by these mirror images could now appear. With this knowledge, they did something most mineralogists didn't do: they squeezed a gem hard to see what would happen. After a few turns of the handle of a vice, the jaws clenched down and Pierre and Jacques found something strange. Surprisingly, the crystal yelped and gave off a small bit of electricity. The Curie brothers discovered that quartz was piezoelectric.

Decades later, Marrison was revisiting quirky quartz; he fashioned it into a small doughnut-shaped slab and zapped it with alternating current. This caused the quartz gem to vibrate steadily, like a piece of Jell-O. Those wiggles could be counted to tell time. But just as a bowl full of Jell-O bounces to its own rhythm, convincing quartz to vibrate accurately was not easy.

By 1927, Marrison had to learn everything about quartz's behavior. His efforts prepared him for the next stage of his work, to create electrical signals to coax quartz to jiggle steadily. While vibrations surrounded Marrison all the time from the rumbling of the elevated train, the buzzing bees from his youth, and his own bellowing voice, he harnessed the vibrations in quartz to do one more thing—tell time. By the end of 1927, Marrison got his quartz clock. This clock employed a quartz ring that was an inch thick and a few inches across in diameter. The clock was so successful that New Yorkers could call the phone number ME7-1212 to get the precise time. A little more than ten years later, time-seeking pedestrians, standing close together but with little interaction with each other, would go to the window display at the corner of Fulton Street in Manhattan to get it.

By the time that New Yorkers made their way over to Marrison's clock, the notion of segmented sleep was a distant memory. The conversion from time taken from natural and biological cues to clock time was complete. Before life was ruled by the clock, segmented sleep was the way of life all over the globe, exercised on several continents. While the way these cultures slept differed, segmented sleep was universal. Since segmented sleep seemed ubiquitous, this raises the question: "Is this the natural way of sleeping?" Anthropologist Matthew Wolf-Meyer, the author of *The Slumbering Masses*, stated that humans "are the only species that seem to consolidate sleep." Researchers found that people living in industrialized cultures, living by the clock, can revert back to segmented sleep. A National Institute of Health (NIH) study by psychiatrist Thomas Wehr plunged seven males into fourteen hours of darkness a day for a month. By the end of the trial, the subjects slept in four-hour segments, with a drowsy state in between. Several researchers and historians believe some of our modern sleep disorders, particularly waking in the middle of the night and the subsequent difficulty returning to sleep, harkens back to the foregone segmented sleep. A. Roger Ekirch, a Virginia Tech professor of history and the author of *At Day's Close,* said that this might be "a remnant of a very strong echo of this older pattern of sleep." What is clear is that the struggle between natural time and clock time resides in our slumber. Our internal sleep clocks differ from the mechanical clocks we obey.

We should sleep better than our ancestors. Yet, fifty to seventy million American suffer from sleep disorders or sleep deprivation.

Nearly one in eight Americans who have trouble sleeping use sleep-aid prescriptions. That number is one in six for those diagnosed with sleep disorders. The National Sleep Foundation implores that we get a minimum of seven hours of continuous sleep, but most Americans get only about six. Poor sleep is not because of our beds. "Our sleep conditions have never been better in all of history," says historian A. Roger Ekirch. Poor sleep seems to be the cost of our inability to get off the clock.

Sleep is a biological necessity. In 1983, scientists illustrated this fact. Researcher Allan Rechtschaffen and coworkers demonstrated the effects of sleep deprivation with rats in a laboratory experiment. In this study, rats were not permitted to sleep, and a range of medical issues emerged, from weakness to poor balance to weight loss to malfunctioning organs. Within fourteen to twenty-one days, many of the rats were dead. In humans, sleep deprivation has been linked to a loss in brain function, obesity, and psychological issues.

Sleep is also cultural. In some countries, naps, siestas, midday breaks, and public napping are part of the social fabric. In contrast, Americans are exhausted, yet unwilling to waste time napping, thanks to our Puritan heritage. Decoupling from time, with a nap or any restorative break, needs a champion for society to embrace it. Although Edison took naps, Churchill took naps, and Einstein did, too, drowsy workers opt to caffeinate. What is clear is that getting some sleep is a deliberate choice. It requires a better understanding of our relationship with time.

For eons, we've looked to make better clocks, but we have lost sleep along the way. What lies at the heart of our sleep dilemma is our culture's view of time. The clock is a yardstick. And for generations society has struggled to make better and better clocks, so we can coordinate our interactions during the day. In our pursuit of better clocks, however, we forgot to look at time itself. That work began around the period of Ruth Belville's business. In another part of Europe the product she sold, time, was being put under the microscope.

Albert and Louis

Uniform timekeeping was always in high demand for office work and for daily affairs, but knowing the precise time was becoming an important need for the biggest business of the day—the railroads. With synchronized clocks, locomotives could run on time, which meant there'd be fewer accidents and more passengers could arrive unharmed.

In 1905, the patent office in Bern, Switzerland, received quite a few applications for inventions with methods to synchronize clocks, particularly for the railroad. To make the synchronization of clocks a reality, eager inventors wrestled with how to set two timepieces that were far apart from each other to the same exact time. Their answer could be the difference between life or death for travelers, and the difference between an unremarkable existence and one of wealth for the inventor. To make sure that the submitted designs were credible, a little-known twenty-six-year-old patent officer acted as a gatekeeper, and dutifully scrutinized whether these inventions were unique, whether their solutions were the cleverest, and whether their embodiments were workable. This patent clerk's efforts could easily have faded into the annals of history, except that his name was Albert Einstein.

Einstein was a bright and precocious young man who wasn't a fan of authority or discipline. He preferred to work alone, so he didn't do well in his studies. After he graduated with a certificate in math instruction, he tried to get a faculty position within the university, but the best employment that young Albert could get was a position at the patent office. For those in the academic ivory tower the patent office was a place for scholarly misfits. Among his peers, Einstein was seen as no Einstein.

Einstein's lowly patent office position was a gift to history, however, for it gave him the space to think; it gave his critical mind a gymnasium in which to exercise; and it propelled Einstein's genius to soar. During the day, he worked on solving real-world problems; at night, he worked on his theories at home. These two disparate activities sharpened each other and honed his skills for seeing things simply.

The railroad had always been the prime mover of timekeeping in the world, and by 1905, when Einstein was pondering it, the railroad completed the last step to converting the world from nature time to clock time. That occasion happened in the United States on November 18, 1883. This day had two noons. Bells tolled twelve times at St. Paul's Chapel near Wall Street in New York City. Then, about four minutes later, the bells rang twelve times again. This was the day that standardized time and time zones were born in the United States and the hours in this country were linked to Greenwich Mean Time in England. The bells marked the death of local time, and the birth of a universal temporal grid for all interactions.

Before the creation of standardized time, regions of the United States were isolated pockets each with its own time zone. Many cities had their own local time based on the position of the sun at high noon. A traveler would find that Michigan had twenty-seven time zones; Indiana, twenty-three; Wisconsin, thirty-nine; and Illinois, twenty-seven. Some train stations had multiple clocks spangled on their walls. In a desire for more uniformity and to reduce confusion, the railroads adopted a standardized time based on Greenwich Mean Time from England. The eight thousand rail stations associated with nearly six hundred independent railroad lines and their fifty-three time schemes were corralled into one system of four time zones. But the train systems, like those in Bern, Switzerland, had the new challenge of determining the time on the train and keeping it in sync with the clocks at the station, which kept Einstein busy in the patent office.

Solutions from inventors poured into the patent office, many of them suggesting an electrical signal or a wireless (radio) signal to send the time. For the idea to be worthy of a patent, Einstein surmised that the best ideas had to meet a certain criterion: one can exchange a signal between clocks if the time for the signal to get from one clock to the other was added to the equation. A rudimentary way to synchronize two stationary clocks was to shoot a flare. But for this to be an effective solution, the time it takes for the flare to reach a certain height must be included. Similarly, more modern methods could use an electrical signal to offer time, but the time it takes for the electrons to travel had to be tallied too. Such means were perfectly fine methods to synchronize clocks and patents along these lines were granted.

But something happened. In Einstein's musings, the problem of synchronizing clocks got complicated when one of the clocks was moving and the time signal was sent by light. Einstein's examination of these patents showed a major flaw not just in the synchronization of time, but in how we think about time itself. And what he discovered would topple our understanding of the physical world.

At the patent office, Einstein distilled how to coordinate a clock at the station with a clock on a train with a simple question: can the period between a tick and tock on a train be seen as the same tick and tock to a person at the station? In 1913, Einstein sketched out his idea for a clock system using light. He determined that if a time signal of light was sent up from the car of a moving train and bounced back downward from a mirror on the train's ceiling, a person on the moving train and another person at the train station would see the light flash differently. Their observation can be likened to how a basketball player sees a basketball while dribbling and how the basketball fans in the bleachers see it. As a basketball player goes down the court, they see the ball bouncing up and down vertically. The same goes for the person on the train, who would see the light signal going up and down vertically, too. Basketball fans sitting in the bleachers, however, see the basketball going up diagonally and coming down diagonally. A person at the train station would see a light signal on the train with a similar path as the basketball—going up at one angle and coming down at another.

The diagonal path is longer than the vertical path. And this is where Einstein rubbed his chin. The speed of light never changes, but one path was longer than the other. To make the unit of time square up between the person on the train and the person at the station, something had to change. To account for the difference, Einstein found that a moving clock is slower than one that is stationary. Time isn't fixed. It stretches.

For generations, scientists such as Sir Isaac Newton believed that time was immutable and unchanging. Newton was of the school of absolutes; Einstein was of the school of relatives. In Einstein's special theory of relativity, our precious unit of time was not the same on one occasion as in the next. A second's duration depended on the speed of the observer.

Humanity preferred certainty in culture and life. However, Einstein uncovered that a second is *not* a second is *not* a second. The time that it takes for a tick followed by a tock will not be identical for a person moving and another person standing firmly on solid ground. Time is elastic. The thing that was so precious for society wasn't exactly what we thought. For generations we have worked to make better clocks from the sun's shadows, to pendulum clocks, to coiled springs, to jiggling gems, and eventually to vibrating atoms in atomic clocks, only to find that what we aimed to measure acts like a rubber band.

Einstein was changing our understanding of time with physics. But just a few years later, in the 1920s, Louis Armstrong altered our experiences with time with music. To many, Armstrong (1901–1971) was a wide-smiling, handkerchief-toting, jazz trumpet player, who crooned "Hello Dolly" and "What a Wonderful World." But Armstrong was much more than the friendly persona that helped his genius navigate an age of Jim Crow. Armstrong was a time traveler, and his vehicle was jazz.

Armstrong came from nothing. He was the grandson of a slave and born in the toughest neighborhoods in New Orleans. According to his biographer, his "small world was bounded at its four corners by the School, the Church, the Dive and the Prison." But, just as he overcame those constraints in his life, he overcame the confines of a musical score. For Armstrong, each eighth note did not need to have the same weight or duration on every occasion that an eighth note appeared. He played them a few hundred milliseconds longer, or shorter, or sooner, or later than what was written on a page. He stretched, squeezed, or shifted notes, giving the music a richness, a feeling, and a forward motion.

Armstrong's attack was a departure from how Western music was usually performed. Western music hinged on accuracy. Marching bands focused on having musicians execute like clockwork. John Philip Sousa, like Sir Isaac Newton, loved precision. Armstrong, like Einstein, found beauty in the lack of it. An eighth note wasn't played exactly as written but swung, and how the note was executed was decided "in the spur of the moment."

Western music and jazz have different approaches to time, fashioned from the cultures from which they originate. In Western music, the notes are an ever-continuous progress to a resounding conclusion; the focus is on the future. In jazz, the focus is on the present. Jazz is an African-American dish that combines European, Caribbean, Afro-Hispanic, and African ingredients. African traditions have a different sense of time. The present is to be savored and expanded. In fact, several African languages have words for "past" and "present" but not for "future." And, it is through this heritage that Armstrong made every note do something, allowing him to stretch the present time with his music.

This African approach to time was transplanted to the New World and took root in the African-American experience. Ralph Ellison captured this black sensibility best in *Invisible Man* when he wrote about the asynchrony of the black experience, existing out of step—before or after—the pulse of time. A listener to Armstrong's work can hear and feel that sentiment embodied in the notes. In Armstrong's *Two Deuces* (1928), he is continuously behind the beat, shadowing it. The notes are delayed and compressed, which creates a gap between Armstrong and his band. To meet up with them again, Armstrong revs up and then hits the gas.

Armstrong stretches not just the notes, but also the listener's sense of time. While the songs on a 78 rpm disc are three short minutes, they are so rich with information it causes our brains to believe the recording is longer than it takes to cook ramen noodles. By playing slower or faster, Armstrong's audience loses track of clock time and the moment speeds up or slows down as they experience it. Einstein showed us that time is relative for the observer; Armstrong made time relative for the listener. How Armstrong shifts our sense of time has been pondered by poets, penned by critics, and probed by musicologists. While the research is still in its infancy, Armstrong's time-shifting abilities might get some support from science.

Timekeeping is ever present in our society. One question that comes to mind is "does timekeeping affect the brain?" The short answers are "yes" and "we don't know." We don't know how the brain was changed as the institution of timekeeping solidified over the span of the nineteenth century, in addition to the loss of segmented sleep. The field of studying the brain's temporal response is fairly new and mostly a twenty-first-century pursuit. What we do know, however, is that the brain gets cues about time from its environment.

Neuroscientists such as David Eagleman have done studies to examine the brain's internal clock. In one experiment, subjects watched a movie with fast-running cheetahs, with their legs coming off the ground, like Trinity in the *Matrix*. During the movie, a red dot of a fixed duration is flashed, when all four legs are suspended in midair. The same experiment is repeated with a small twist; in the second experiment, the same cheetah movie was played in slow motion, and the same annoying red dot blinked for the same duration as it had before when the cheetah caught air at normal speed. After the tests were compared, moviegoers believed that the red dot during the slow motion movie was shorter. "Your brain says I need to readjust my sense of time," said Eagleman. Our brain determines the time based on our knowledge of the laws of physics. Our perception of time is shaped by the events it uses to measure time—the landing of a wildcat's paw or perhaps the duration of an eighth note.

On a personal level, we have always been aware of the elasticity of time. Good times seem short and bad times seem to last forever. Neuroscientists have shown that in some respects this is not fiction. The length of our memories is linked to how good or bad the occasions are. What neuroscientists have found is that we don't perceive time slowing down in the moment, *but* our recall of the event makes us believe that time has slowed down. To understand what is going on in the brain, imagine that the brain acts like a computer that stores information on a hard drive. When life is boring, the hard drive stores a regular amount of information. When we are scared, however, as during a car accident, the brain's amygdala—our internal emergency operator—kicks in. Our brains collect finer details like the crumpling of the hood, the breaking off of side-view mirrors, and the changing of the expression on the other driver's face. The amount of detail gathered is increased, as if two hard drives are storing the data. "You're laying down memories on a secondary memory system now, not just one," said Eagleman.

More data gets stored. When the brain recalls the event, it interprets the large amount of information as a longer incident. The shape of the memory becomes the brain's yardstick of time.

Science shows that the size of a memory and our perception of time are coupled like gear teeth in a bicycle chain. Rich and novel experiences, like the recollections of the summers of our youth, have lots of new information associated with them. During those hot days, we learned how to swim or traveled to new places or mastered riding a bike without training wheels. The days went by slowly with those adventures. Yet, our adult lives have less novelty and newness, and are full of repeated tasks such as commuting or sending email or doing paperwork. The associated information filed for those chores is smaller and there is less new footage for the recall part of the brain to draw upon. Our brain interprets these days filled with boring events as shorter, so summers swiftly speed by.

Despite our desire for better clocks, our measuring stick of time isn't fixed. We don't measure time with seconds, like our clocks, but by our experiences. For us, time can slow down or time can fly.

For ages, humans have had an evolving obsession with time. Time helped us understand the world, helped us make appointments, and helped us interact. In our pursuit for precise timepieces, we abandoned nature's cues, of sunrises and sunsets—and we lost sleep—in hopes of possessing time with great precision in timekeeping. But time is not something that can be possessed. Einstein showed us that time is elastic, and what time it is depends on who you ask. Armstrong demonstrated our brains are faulty clocks, speeding up and slowing down with outward cues. But both Einstein and Armstrong, using science and jazz, showed that we are the time we keep.

For nearly half a century, Ruth Belville persevered and provided time to customers in London. Her work was considered stuck in time, especially by the businessmen who tried to poach her customers with their telegraph clock service. But the older technology of Arnold the watch was accurate to a tenth of a second, while electronic pulses had accuracies of one second. Ruth also brought something to her customers that the metal wires of the telegraph could not bring. For an annual fee of £4 and an occasional spot of tea, she brought a touch of humanity to those with whom she interacted, sharing a bit of banter and news, which she collected on her journey. Eventually, however, time services using technologies of the telegraph, wireless, and radio slowly and gradually shrunk her business to about fifty clients from the one hundred had by her mother and the two hundred by her father.

After decades of service of bringing the time to Londoners, Ruth retired. In 1943, Ruth Belville, the Greenwich Time Lady, died accidentally in her sleep when she asphyxiated on carbon monoxide leaking from her gas lamp turned to a low setting. On her nightstand was Arnold, her trusted companion, which stopped ticking a few days later. Her passing was the end of a century-old time distribution service. Together with Arnold, Ruth provided the time, but in the end she had a limited supply of it.

2

Connect

How steel stitched the country together with rails, but also how steel helped to manufacture culture.

The Connector

In the early hours of Friday, April 21, 1865, bodies began to pack into the streets of downtown Baltimore. As sunlight broke through the light rain, the thick gathering of humanity near Camden Street train station made the roads impassable. Work stopped. Schools closed. Stores emptied. The people, waiting eagerly for the arrival of a train, wept.

That anticipated train chugged into the station containing the remains of President Abraham Lincoln. He died on April 15, a few days after the end of the Civil War. Now, inside this train, dubbed "The Lincoln Special," the late president's body sat dressed in the same suit he wore six weeks earlier at his second inauguration.

The grief-stricken public begged that Lincoln's funeral services be extended beyond Washington. In an age before television or radio, the only way for a person to participate in his memorial service was to leave her farm or close his shop and travel to go where Lincoln lay in state. Lincoln's funeral train allowed the nation to mourn in unison by bringing him to them in a way that neither a telegraph nor newspaper could satisfy. Over the course of thirteen days, the train sojourned from Washington to Baltimore to Harrisburg to Philadelphia to New York to Albany to Buffalo to Cleveland to Columbus to Indianapolis to Chicago, stopping at each, before his final destination of Springfield, Illinois, for his burial.

The month of April 1865 was one of the most tumultuous in American history. The good news of the end of the Civil War and Ulysses S. Grant's conquest of Richmond on April 9 flooded the country. Church bells tolled. Fireworks boomed. Revelers cheered. Less than a week later, however, those celebrations were silenced by the sad news of Lincoln's assassination.

Coordinating the transportation of Lincoln's remains fell on the shoulders of the secretary of war, Edwin Stanton. His temperament was the opposite of Lincoln's, but he kept a loyal vigil at Lincoln's deathbed and took on the challenge of creating the biggest funeral that the country had ever seen. Stanton made this procession possible by designating the railroads as military domains, so the railroad companies that ran individual lines had to fully cooperate.

Orchestrating the fifteen railroad companies was a mammoth undertaking. For that reason, Stanton created the Committee of Arrangements and gave it full power to make this funeral procession possible. Its members were "authorized to arrange the time tables with the respective railroad companies, and do and regulate all things for safe and appropriate transportation." While railroads were the circulatory system of the nation, the nation was fragmented. Scheduling the train required a wrangling of various time zones between cities and states, which were more numerous and less systematic than today. Before standardized time began in 1883, most towns told time by high noon. Clocks had to be increased one minute for every twelve miles moving east to keep the correct time. Noon in DC was 12:12 in New York, 11:17 in Chicago, and 12:07

in Philadelphia. As such, the country was composed of nonunified regions—splintered by both war and local time—and this train, with its precious cargo, briefly threaded them together.

Lincoln's funeral car was a magnificent carriage. The sides of the car were painted a rich brownish-red, which was lovingly hand-polished to a shiny gloss with oil and rottenstone. The interior was lathered with plush green upholstered walls and black walnut molding. Pale green silk curtains cascaded along etched glass windows and three oil lamps beaconed from the car at night. With sixteen wheels instead of eight, it was a vehicle that paralleled those of European royalty. This special coach had three stately compartments and Lincoln's casket sat in the end room. As a presidential car, it was designed to be Lincoln's Air Force One. But here, festooned in black bunting and on its maiden voyage, it was now Lincoln's hearse.

At each destination along the route, the train stopped and the honor guards, dressed in robin's egg blue uniforms, carried Lincoln's body in huge formal processions to viewing areas. Oceans of people waited for hours and many watched processions from windows or roofs or trees. In the halls, thousands of mourners, sometimes packed twelve bodies deep, stood weeping and waiting for a glimpse of the open coffin. For many, this was the very first time they were seeing Lincoln's face, since photographs in newspapers were still rare.

As Lincoln grew closer to his resting place, the emotions of the nation heightened. In some whistle-stops, the number of mourners was greater than the town's population. Crowds of those who could not travel to cities made their way to the train's tracks.

The locomotive engine, bearing a portrait of Lincoln in the front, chuffed at a cautious twenty miles an hour and passed by train stations at five miles an hour. There were nine cars consisting of six passenger and baggage cars, a car for guards, a special car that carried the body, and the last car for family and the honor guards.

A pilot train ran ten minutes ahead of the funeral train and announced Lincoln's arrival by sounding a half-muffled bell, with a leather pad placed on part of the bell's clapper to soften the strike. Those waiting trackside heard a rhythmic ringing of a clear tone followed by its echo and knew it was time to prepare. In an age before Edison's electric lights, bonfires were lit at night and set along the train's route to push back the darkness.

There was a solemn eagerness as people lined the tracks day and night. At the sight of the train, they stood back—some waving small flags, some standing silently, some singing hymns. Then, fifteen minutes later, another train followed. When this train passed, the crowds stepped into the tracks and watched it fade into the distance. Then, the moment was over.

Before Lincoln was set in his final resting place, his remains would be transported across the country on over sixteen hundred miles of track. Millions of people participated. Nearly every American knew someone who attended memorial services or watched funeral processions or saw the train pass by. In these sad and dark days, the nation was stitched together with iron rails. But soon, those rails would turn to steel—once the secret for making lots of it was unlocked—and do so much more.

Steel, a metal alloy that is hidden in plain sight, would be the great connector of the country (an Abraham Lincoln in a way, but in material form). In order for steel to bridge and link the United States together, though, the recipe for making tons of it rapidly had to be puzzled out. That work would be the endeavor of an inventor in England who would be unable to forecast what customs his creation would create.

Bessemer's Volcano

Henry Bessemer dreamed about steel. He longed to make an unlimited supply of it. Although in 1855 he didn't know much about the science of steel or the recipes for steelmaking, that didn't stop him from trying. It never had.

Bessemer was a prolific English inventor with over one hundred patents to his credit. His most famous invention to date was a gold-colored paint that didn't contain any gold. In the 1840s, metallic paint was the must-have item in England and used to gild ordinary frames, converting them into ornate ones. When buying this paint as a gift for his sister, he was flabbergasted when he found it cost as much as a laborer's day wage. So he figured out how to machine bronze into a powder that glittered like gold at a fraction of its cost. When he added this powder into paint, he made an inexpensive alternative that anyone could buy. And everyone did, making him rich. But soon Bessemer's thoughts turned away from gold and its glitter for decorations and toward steel with its strength for weapons. Little did he know that his imaginings for making steel would take him on a journey that would change the world, too.

In 1853, England and its allies (France, Turkey, and Sardinia) were at war in a conflict known today as the Crimean War—a battle to let Catholic pilgrims have access to the Holy Land. The allies supported the Catholics; the Russians didn't and wanted to preserve the Holy Land for Orthodox Christians. This tension ignited into combat and many inventors, like Bessemer, focused on making better weapons for the military.

To win the war, England needed steel—and lots of it. Steel is a strong metal that could build powerful cannons. Unfortunately, the creation of certain types of steel, like blister steel, proceeded at a glacial pace, while other processes, like making crucible steel, were difficult to scale up. In 1855, two years into the war, it became clear that the inventor who found a way to make steel quickly and cheaply would become very rich. For entrepreneurs like Bessemer, steel could lead to an economic alchemy. Better steel for cannons would turn to more gold in his pockets.

Bessemer's path to becoming an inventor was no accident, but the design of his father, Anthony Bessemer. The elder Bessemer was a Londoner working in Paris. A respected inventor himself, at the age of twenty-five he was elected to the lauded French Academy of Sciences for his typesetting devices and for his improvements to the optical microscope. Anthony's path crossed those of scientific elites like Antoine Lavoisier, who was the discoverer of oxygen, and often called the father of modern chemistry for his system of chemical names. Anthony had the golden touch for inventing and seemed to do no wrong. All that came to an end, however, with the French Revolution in 1792. Under the bloody Reign of Terror, its architect Robespierre wanted to build a republic; he had no tolerance for the monarchy or the sciences and took to eradicating them both. Under Robespierre's reign, Anthony and other scientists weren't safe. So Anthony hastened back to England penniless, barely escaping the guillotine—unlike Lavoisier. Settling in a quiet and small town in England, Anthony rebuilt his typesetting workshop and focused his energies on his best invention yet: his son Henry.

Henry Bessemer was born in Charlton, England, in 1813. He had very little formal education, but he was allowed free rein in his father's workshop. There, he received tools instead of toys, and his desire to build was nurtured. Henry matured into a tall and barrel-chested man, with a bold nose, meaty jowls, and thick sideburns that imperfectly distracted from the lack of hair on the top of his head.

Like many brilliant people, Bessemer was a study in contradiction: He was at times engaging, but at other times explosive. He was stubborn but spontaneous, generous but overbearing. He was talkative, but he preferred solitude with his machines. Even his physique presented a paradox: his chest was stout, but his legs were lanky. And, while Bessemer's eyes sometimes looked sad and preoccupied, he was always looking for new opportunities. And here, in his early forties, his moment had come—the mission of producing steel cheaply, rapidly, and abundantly.

Bessemer embarked on making steel, which could be defined as iron with a bit of carbon in it. But that definition doesn't adequately describe the marvelous metamorphosis that occurs when carbon combines with iron. On the microscopic scale, part of steel curiously and simultaneously turns into two different substances that stack into multiple layers, similar to those of a cake. One of the layers is rich with carbon; the other is not. One of the layers is extremely hard; the other is not. These layers complement each other with strength and malleability (the ability to be bent). Being strong and malleable usually doesn't happen at the same time in a metal. These characteristics are more often like two ends of a seesaw—as one goes up, the other goes down. But in steel, both exist, because the layers in steel have each quality. These two layers with opposite properties make steel versatile.

This mysterious marriage of carbon in iron gave rise to strong steel, which can be used in making durable cannons. But creating steel wasn't going to be easy for Bessemer. Adding the perfect amount of carbon into iron was a page torn out of the classic story "Goldilocks and the Three Bears." Too little carbon makes the steel too soft. Too much carbon, say over 2 percent, compels steel to break like chalk, making cannons dangerous—not for the intended target, but for the person shooting, since cannons made of brittle metals can explode. A steel that was "just right" for cannons meant that a specific percentage of carbon needed to be added into the iron at an amount of 1 percent or less, and this process needed to be done properly over and over again.

Bessemer understood this and the problem was even more compli-cated because pure iron was not the starting point. In addition, the structural metals that were readily available at the time were cast iron and wrought iron. While their names have the word iron in them, they also contained ingredients that made them undesirable for Bessemer's specific needs. Cast iron was a combination of iron and carbon with too much carbon in it, so it was brittle. Addition-ally, it could not be welded or forged into a cannon shape. On the other hand, wrought iron contained nearly no carbon, so it could be worked—into plates for ship hulls, for example—but it was often full of unwanted impurities, known as slag, which jeopardized the cannon's strength. While Goldilocks had the choice of porridge that was too hot and too cold, Bessemer's choices were metals that were too brittle and too soft.

Bessemer's thinking was that steel could be made somehow by first removing carbon from a carbon-rich crude iron called pig iron. With this new method, he could usher humanity into a new age of steel if he just found the trick. As he attempted do so, he made "many modifications and alterations" to his furnace to get the right conditions. Being single-minded in solving problems was one of Bessemer's best traits. He became obsessed by metal and an idea came to him while he was recovering from being very ill.

Bessemer was a man of steel, but he also had his Kryptonite. He had an unusual eccentricity of being prone to acute attacks of sea-sickness. His bouts flattened him for days. "Few persons have suf-fered more severely than I have from seasickness," he recounted in his autobiography. While convalescing after a long ship voyage, he had a Eureka moment. To burn off carbon in the iron, he was going to need air—and lots of it. He wrote, "I became convinced that if air could be brought into contact with a sufficiently extensive surface of molten crude iron [pig iron], it would rapidly convert it into mal-leable iron."

The blowing of air is an old technique used by expert tailgaters, BBQ pit masters, and prehistoric people to make fires hotter. Bessemer copied this idea in 1855, but put a twist on it, using the air for a slightly different reason. He was going to use air to chemically combine with the carbon in the molten pig iron as a way to remove the excess carbon, so he could add a precise amount back to make steel. Bessemer blew the air directly into the molten metal bath from a pipe inserted at the bottom, which harkened back to the behavior of a volcano. It was a crazy idea, but it worked.

He described his experiment the following way: "All went on quietly for about ten minutes." Occasionally he would see sparks, but that did not alarm him since that was to be expected when forcing air into molten metal. He anticipated a bubbling cauldron with fire and smoke. But a few minutes later that fire and smoke changed to an inferno. The oxygen from the air chemically—and violently—reacted with the carbon and "sent up an ever-increasing stream of sparks and a voluminous white flame," followed by a battery of loud explosions. When the oxygen and carbon chemically combined, his nose, ears, eyes, and skin were assaulted by the reaction's thick smoke, thunderous pops, bright flames, and scorching heat. This bubbling demonic brew of molten metal became Vesuvius.

While Bessemer described what happened in an unemotional way, what is clear is that his experiment erupted and burnt part of the building's roof. After the fires died down and the debris was cleared, he found he had been successful. To his delight, this chemical eruption removed the carbon from the iron, which he was able to add back in desirable amounts to make steel.

After years of work to perfect his recipe, Bessemer got his steel, but it was too late to be useful for the military. The war was over and the Russians got a shellacking without it. But being a resilient entrepreneur, Bessemer followed his personal motto of "Onward Ever" and set his sights on a new and promising market—the railroads.

The Virtual Man of Steel

When the news of Sir Henry Bessemer's process of shooting air into a vat of iron to make steel reached the United States in the fall of 1856, the announcement filled the nation with glee. Steel would connect the nation with bridges and stitch it together with rails. The news about Bessemer's invention filled William Kelly with dread, however. Kelly also had an air-blowing recipe for making metal, which in his mind seemed similar to Bessemer's idea. If Kelly wanted to leave his mark, he was going to have to beat Bessemer and get an application to the patent office swiftly.

All William Kelly wanted in life was to be significant like his father. His father was a well-respected and wealthy city elder of Pittsburgh, but the younger Kelly did not inherit those traits for success. William Kelly, born in 1811, grew to be a tall, thin man, who seemed short on ambition. He entered the clothing business and was a traveling salesman with his brother, John, in a company called McShane & Kelly. It was good work, he got to see the country, and he was a senior partner. But fate had other plans. A fire burned down his company's warehouse. Around the same time, William met Mildred Gracy in a town not far from Cincinnati called Eddyville, Kentucky, on one of his many business trips. He moved there to be closer to her.

Kelly started all over again in his late thirties, living in a tightly knit community as a stranger in a very strange land. To earn a living in this small rural town, he bought the Eddyville Iron Works with his brother in 1847, which they renamed Kelly & Company. William Kelly married Mildred and got additional financial backing for his ironworks from his wealthy father-in-law. The ironworks was located on the banks of the Cumberland River and had two operations that were separated by a few miles: the Suwanee Furnace and Union Forge. The furnace converted iron ore from the mines into pig iron, and the finery forge converted that metal into wrought iron bars. William managed the furnace and forge operations, while his brother handled the finances. Neither brother had experience in iron making.

The company had all the equipment to transform pig iron to wrought iron, that is, from iron with lots of carbon in it to one with less. Iron with less carbon, say less than 0.4 percent, is desirable because it is strong and not prone to cracking. Pig iron, on the other hand, has over 4 percent carbon.

Kelly's ironworks was positioned to do well and had plenty of resources to keep it going; in particular, it had a good source of iron ore and had large landholdings of timberland nearby. The wood would be converted into charcoal that was used to keep the furnaces hot. Fuel for the furnace was one of the biggest costs of running an ironworks, so Kelly wanted to find a way to keep the operation running economically.

The legend goes that one day in 1847, Kelly saw one of his workers at the refinery blow air onto the surface of a molten vat of pig iron. As an observant novice, he expected that the air would cool the metal, but the air blast did the opposite. The molten pool got hotter. Adding air to this molten metal increased the temperature because a chemical reaction occurred. "After close observation," Kelly wrote years later, "I conceived the idea that after the metal was melted, the use of fuel would be unnecessary." Kelly observed that the blast of air would raise the temperature and reduce the need for more wood to keep the furnace's flames going. He looked at this air-blowing process as a way to save fuel.

Unbeknownst to Kelly, the air did more than that. Kelly's air blowing process, or pneumatic process, as he called it, removed carbon, making the molten iron a good starting point to make steel. Steel could be made by adding a specific amount of carbon back into it. Kelly had created something significant by blowing air onto the melt, but he didn't know what that was. Not yet.

News stories of Bessemer's impending American patent swelled; his US patent application was submitted in 1856. Bessemer had a process of blowing air into a molten metal. This process seemed similar to what Kelly conceived of as his process, but Bessemer's reasons for blowing air were different. Bessemer had knowledge that blowing air into the molten pool chemically removed carbon, which he could add back in precise amounts to make an ideal steel. Kelly's understanding was that blowing air reduced the need for fuel.

Kelly brought a case of interference on September 30, 1856, to the United States Patent Office, just a few weeks after learning of Bessemer's work. Kelly sought priority for his invention, which he performed in 1847. He produced over a dozen witnesses, and defeated Bessemer's application.

But the difference between Bessemer and Kelly is that Bessemer had a process that worked; Kelly didn't. Blowing air was not the only action needed to bring iron closer to creating steel. Bessemer had learned this the hard way.

Bessemer's early experiments removed carbon from pig iron, just like Kelly's pneumatic process. This is a good first step because too much carbon makes steel brittle, and it will snap like a raw carrot stick. But making great steel requires attention to other ingredients, namely phosphorus and manganese. If steel has too much phosphorus, it will be brittle, so removing phosphorus is a good thing. Manganese has the opposite affect, however: steel with too little manganese will also be brittle. Making steel is unforgiving; it is a metallurgical soufflé.

Bessemer's initial process inadvertently took out manganese, but it did not remove phosphorous. He had unknowingly used pig iron that did not have much phosphorous in it to begin with. He was lucky, but those who tried to repeat his experiments would not be so fortunate. What they created was inferior, and not what he had advertised. So, when Bessemer sold licenses, he made a fortune, but soon had to return all that money and pay fines when he was sued. Eventually, Bessemer had to combine his patent with the work of Robert Mushet, who had the patent for adding manganese, and Sidney Thomas, who had the patent for removing phosphorus. And together this process, which was called the Bessemer process, worked. While there was a debate about all the chemical reactions happening, there was no debating that Kelly was ignorant of them. Judging by the testimony from the patent office, it seems that Kelly's men understood the introduction of air across the molten metal as a way to reduce fuel and not as a way to create a superior metal.

The patent office heard Kelly's case, ignored Kelly's lack of scientific understanding, and overlooked the inconsistencies between his patent claims and his evidence. The American patent office let the American inventor get the American patent. On June 23, 1857, patent #17,628 was issued, specifying that air increases the heat of the melt "without the use of fuel." The patent's title, "Improvement in the Manufacture of Iron," makes no mention of steel.

Although he had this patent in his possession, Kelly didn't do much with it. There is little evidence of continued steel-making work and no reference to it in his letters. Kelly also went bankrupt. After the Panic of 1857, when the downturn in Britain reached the banks in New York and soon the rest of the United States, Kelly could not get financial backing. He had to close up shop. He would never be significant and the mass production of steel would have to wait. The development of railroads and bridges rested in the hands of a man who had the patent but made little effort to manufacture steel. That wait would be prolonged for years when the United States Patent Office renewed Kelly's application, rejecting Bessemer's again.

Soon, American industrialists became impatient as the need for steel swelled with the start of the Civil War. Rails made solely of iron only lasted two years and therefore needed to be replaced frequently. Yet steel rails lasted eighteen years. US companies sought to license the steel-making process. In the end, a legal agreement was reached and all the steps to make steel rapidly were combined. The removal of carbon with the air-blowing process was applied along with the steps of adding manganese, removing phosphorus, and then adding precise amounts of carbon back in.

Many would say that Bessemer won. He did. The steel making process was known all over the United States as the Bessemer process, and Henry Bessemer became richer. But Kelly won, too, by gaining the significance he long craved. In a town not far from Eddyville, Kentucky, a marker for the Kelly Furnace reads: "Here William Kelly (1811–1888) discovered a steel making method later known as the Bessemer Process, which made it possible for civilization to pass from the Iron Age to the Steel Age."

As the manufacturing of steel became a reality, this great material grew in volumes to build a nation, but in the course of creating steel, a legend was swirled into its recipe, too.

How Steel Changed Us

The Bessemer process conjures up the image of a volcano in a cauldron. The fantastically hot temperatures create a molten mixture of iron and carbon that glows bright orange, and the super-heated air around it undulates nearby images in and out of focus. Looking into the mouth of the brew, a calm haze sits on the bubbling surface with fingers of flames darting out without rhythm or intention. Billows of smoke glide up from the surface and are dotted with sharp sparks of yellow and orange. But the smoke, flames, and sparks are just the opening act. Air is forced into the crucible to create a cocktail of forest fire with a dash of the Fourth of July. The bubbling blend thunders and the carbon and air are devoured by the melt. The eyes are assaulted as the colors change from red to orange to yellow to a blinding white. The brew of molten metal has been transformed. This is the birth of steel and the world we know.

This molten brew gave rise to the steel rails. The steel rails became a web and the connective tissue for the country. And with that, many things emerged. As one can imagine, people began to migrate and cities grew. Take Chicago, for example: this railroad hub swelled. In 1850, it had 30,000 people; by 1890, the city had tripled in size. Not only did cities grow; cities that did not exist before were born. A host of dusty towns peppered along the rail lines became the full-blown cities known today. Metropolises like Albuquerque, Atlanta, Billings, Cheyenne, Fresno, Reno, Riverside, Tacoma, and Tucson are successful offspring of the rails. The rails had such a hold on life: if you were connected, you thrived; if you were not connected, you might not survive.

Travel before the railroads is unfathomable to our modern imagination. A sense of a stagecoach journey was provided by Josiah Quincy (1772–1864), the fifteenth president of Harvard, when he shared details about his excursion from Boston to New York:

> The journey to New York took up a week. The carriages were old and shackling and much of the harness was made of ropes. One pair of horses carried the stage eighteen miles. We generally reached our resting place for the night, if no accident intervened, at ten o'clock and after a frugal supper went to bed with a notice that we should be called at three the next morning, which generally proved to be half past two. Then, whether it snowed or rained, the traveler must rise and make ready by the help of a horn lantern and a farthing candle, and proceed on his way over bad roads, sometimes with a driver showing no doubtful symptoms of drunkenness, which good-hearted passengers never fail to improve at every stopping place by urging upon him another glass of toddy. Thus we traveled eighteen miles a stage, sometimes obliged to get out and help the coachman lift the coach out of a quagmire or rut, and arrived at New York after a week's hard traveling.

Traveling by stagecoach was part expedition, part rock tumbler, which made the train a welcome guest. And with the ease of traveling with the railroads, maps could be mentally redrawn. An example of this recasting of distance can be seen in the *Atlas of the Geography of the Unites States* from 1932, which culls census data about population and demographics, as well as how long it takes to travel from one point to another, or the rate of travel. (See figures 18 and 19 for details.) These rates of travel are depicted on the maps with curves that resemble the contour lines found on serious hiking maps to show elevation. Starting from New York City, the map shows how far one can travel over specific time periods. As the map depicts, traveling from New York to Washington, DC, in the early 1800s would take five days by stagecoach. The contour lines widen for trips taken just a few decades later, however. In the mid-1800s, that journey from New York City to DC took only a day by train. Before the rails, if a son moved his nuclear family fifty miles from his childhood home, that visit would take two days and would be infrequent; with the rails, that same trip would take two hours and grandmothers would get to see their grandchildren. With the railroads, the nation experienced what geographers call *time-space compression*. That is, as the time that it takes to get from one place to another decreased, the significance of the distance between these points decreased, too. In other words, the world shrank.

Prior to the railroads, going twenty to thirty miles per hour was breathtakingly fast; this was two to three times the speed of a stagecoach. As with anything new, there was resistance. "If God had designed that His intelligent creatures should travel at the frightful speed of 15 miles an hour, by steam [railroad], He would clearly have foretold it through His holy prophets." This was the sentiment of the Lancaster, Ohio school board in 1828 on the coming of the railroads to connect the west to the Mississippi River. But trains came despite the disapproval and with them came speed.

With the steel rails, the nature of commerce itself was altered, too. Before the trains, small stores had to hold on to large inventories, which came with some risk of damage or theft. Railroads could bring in new items and replenish inventories every few weeks, allowing store owners to operate with smaller supplies and less risk. Additionally, railroads changed the nature of small businesses. Before railroads, merchants on the western frontier worked seasonally. In the summer, sales were steady, but with the winters came frozen canals and rivers preventing patrons and products from reaching the store. Commerce back then was a feast followed by a famine. Railroads brought a steady stream of goods, since travel was divorced and disconnected from nature's winter freezes.

That molten brew of iron and carbon that made steel rails possible helped diffuse products across the nation, moving us away from considering only goods that were locally available. The rails grew at a precipitous rate in the United States. In 1840, before the Bessemer process, there were 3,326 miles of railroad tracks. By 1860, just twenty years later, there were 30,600 miles—a little more than the distance around the equator. By 1900, there were enough steel rails to go around the world ten times. This means almost any corner of the United States was now connected and with that access came the nation's appetite for that region's products.

Steel made dinnertime more satisfying to the country's palette. To get great quantities of food to everyone's plate, the rails connected regions with the supply of products to meet the nation's demand. Before railroads, communities subsisted exclusively on their own labor, where buying and shopping locally was just the way of life. But the rails connected the country to other products, changing that thinking. Minneapolis was rich with flour; Chicago was rich in cattle; Louisiana was rich in sugar; Missouri was rich in corn. Each of these regions was willing to supply others with their products for items that they needed. For exchanges of these resources to occur, a cheap means of transportation was required, and that came in the form of the railroads.

A brew of iron and carbon created steel and rails of steel criss-crossed the country, helping to build—and feed—the nation. However, steel was not quite done. Not yet.

An Amalgamated Holiday

The Christmas holiday as we now know it didn't always exist that way. Santa and his reindeer didn't appear a year after the birth of Christ; children would have to wait over three centuries, at least, for Christmas to come. By the 1800s, this holiday had picked up and combined various elements of European religious and pagan traditions along the way, taking the form recognizable to our modern eyes with the publication of Charles Dickens's *A Christmas Carol* in England in 1843. With Dickens's book—and its perennially recognized characters of Ebenezer Scrooge and Tiny Tim—the final wrapping of this winter holiday tradition was complete.

Despite the enthusiasm for Christmas abroad, it wasn't all that popular when it arrived in the United States. "They made more of New Year's than of Christmas," said one-hundred-and-one-year-old Mrs. Jane Ann Brown to the *New York Times* in 1894. Mrs. Brown had seen this holiday change over the course of her long life. What she saw was that, unlike today, New Yorkers didn't think much of Christmas. They weren't alone.

In Philadelphia, Christmas began as an unsavory holiday during which intoxicated revelers took to the streets, begging for money. Factories were shut for the winter, leaving workers out of funds, so these unemployed souls used this holiday to go to the doors of the wealthy to perform for donations. This down-and-out custom of singing for alms eventually evolved into the much-more-delightful act of caroling. Christmas, in general, was then elevated and shaped to align with middle-class values, becoming a time of "gift-giving, stocking-filling, and a family dinner," as Professor Susan Davis has written.

Christmas's transformation is evidenced by the number of Christmas carols created during this time period, which include:

1839   Joy to the World

1840   Hark! The Herald Angels Sing

1847   O Holy Night

1850   It Came upon a Midnight Clear

1857   We Three Kings

1857   Jingle Bells

1868   O Little Town of Bethlehem

But there is a dark underbelly to this holiday's metamorphosis, too. Historian Penne Restad states that Christmas was transformed into a gift-giving occasion to keep the economy going. And the best way to move products, presents, and Christmas items was on rails of steel.

The pieces of Christmas were stitched together into a mammoth tapestry. First came the Christmas tree, the selling of which became a brisk business in the nineteenth century. "There is now in progress a market where competition is as sharp and dealing as close as can be found in any mercantile exchanges," said the *New York Times* in 1893, "this market deals only in Christmas trees." From early December until Christmas day, dealers from Maine sold trees in cities. While this is not uncommon today, selling Christmas trees was newsworthy back then. These Christmas trees were hauled by trains from Maine to New York City on rails of steel.

Along with the Christmas tree came the Christmas card. "Why four years ago a Christmas card was a rare thing," said one postal official in 1882. "The public then got the mania and the business seems to be getting larger every year." The last ingredient in the holiday concoction was the tradition of giving gifts. The *New York Times* stated in 1890, that there is "an epidemic of giving and receiving presents." Not everyone was happy with how Christmas was going. "It seems the fashion to be extravagant, almost reckless, in expenditure," said the *New York Times* in 1880, "people of all classes vie with each other in the costliness of their presents." Despite these virtuous sentiments, society was swept up by the momentum propelled by trains full of Christmas trees, Christmas cards, and Christmas gifts moving on rails of steel.

Some scholars will claim that the United States was fragmented after the Civil War and the death of Lincoln, and that there was a need for a great unifier. This winter holiday was hatched to serve as a connector of the country. Using the rails, business and railroads stitched this Christmas creation together. Shopping was part of the American culture, and the steel rails enabled shopping. The trains brought in the products and the trains brought people to the stores to buy these products, creating a circulatory system. Christmas provided the rapid pulse.

The Christmas we know was born in a boardroom, swaddled in steel. Further evidence of how commerce manufactured Christmas comes from gazing at the location of Thanksgiving on the calendar. Abraham Lincoln declared it a national holiday to occur on the last Thursday in November. Decades later, Franklin D. Roosevelt, on the urging of business leaders and department store lobbyists, moved Thanksgiving back another week to the third Thursday of November. This shift lengthened the Christmas season and gave people more time to shop. With the stroke of the presidential pen, the trains puffed across rails, happy to bring Santa's gifts to little girls and boys and help them enjoy the holiday they—and steel—helped to create.

In 1884, *Science* magazine stated that Henry Bessemer's process was "an invention which has, in the short space of a quarter of a century,

completely revolutionized some of the greatest of human industries." Once steel became plentiful, it was used in railroad lines to
link the country together. The country needed a powerful connector, and the first instance came in the form of a great man named
Lincoln. A material named steel would also connect the country,
but in a different way. Bessemer's molten mixture of carbon and
iron first compressed space and then constructed many curious creations from cities to commerce to the Christmas we know, thrusting
our society into this peculiar and complex age.

3
Convey

How telegraph wires of iron and later copper gave rise to rapid forms of communication, and how these wires shaped information—and meaning.

By the Dawn's Early Light

On an early January morning in 1815, Major General Andrew Jackson gazed through his spyglass toward the Louisiana battlefield where he had been holding back the British for weeks. Standing on the muddy banks of the Mississippi, just six miles south of the city of New Orleans, the men enlisted in his troops were no professionals by any stretch of the imagination. Only a fraction of them had been trained as soldiers, and the rest were frontiersman, volunteers, businessmen, freed slaves, Native Americans, and a few pirates. America's existence rested in the hands of four thousand ill-prepared men opposing the British army's trained ten thousand.

The War of 1812 had been well under way for three years, and the news about its progress, traveling by horse or by boat, took weeks to reach the public. As to be expected, the new nation of America, which hadn't built a sizable army yet, wasn't winning. Additionally, there was strife outside of the nation as well as within it. The states were not united in the north, and the British were certain that a blow to the south in New Orleans would also reverberate to the west, causing this nation to collapse. The only person standing in their way of succeeding was a pugnacious, hot-tempered general named Andrew Jackson.

The battle at New Orleans was between enemies, but also between temperaments. Leading the British was Major General Sir Edward Pakenham, a thirty-six-year-old educated military man with family ties to royalty. Young, strapping, and dashing, he was called Ned, for his ease among his soldiers, who highly regarded him. Jackson, on the other hand, was forty-seven. Haggard and drawn by the dysentery riddling his body, a bullet sat near his heart, lodged in his chest from a pistol duel from 1805. His troops had a moniker for him too; they called him Old Hickory, for it was the hardest wood they knew. Jackson had little education, was an untested commander, and was no military genius, but he possessed the fierceness of a Rottweiler.

By the dawn's early light, on January 8, 1815, Britain's third attempt to mow down the American troops commenced. The battlefield near New Orleans was in the fields of a sugar plantation, hemmed in between the brown waters of the Mississippi and a black swamp. Jackson's line was a rampart with a moat in front that was strong in the center, but weaker on its ends. Their enemy created a plan to attack each side, to take advantage of this frailty. Pakenham, the commander of the British, was a thinking man who employed a complicated battle strategy, requiring a precise coordination of movements, functioning together perfectly like the parts of a clock. There were a total of four thrusts: one across and up the Mississippi, one up the river's edge, another up the swamp's edge, and the last charging right up the middle. Pakenham saw this battlefield as a cerebral contest of chess, winning by capturing a single piece;

however, Jackson saw this as a down-home game of checkers, winning with more pieces on the board.

As shots were fired, nature tipped the scale in favor of the Americans. British boats sent up the Mississippi River got stuck in the mud and their attack from the back of the Jackson line came too late. Human failings played a role, too. The plan was for another British thrust to charge on foot, carrying ladders to climb the rampart as well as bundles of sugarcane to fill the trenches, allowing them to march over the water. But the ladders and bundles were forgotten. When this "most extraordinary blunder" was realized, they retreated to retrieve these items, and then returned, further slipping the timing of chess moves, disrupting the synchronization of the campaign.

As red rockets blared over the Americans, the blasts stirred up trepidation in the un-battleworn. Jackson steadied them, like unbroken steeds, with firm, unfazed, and fatherly words, overriding their impulses to flee. On his command, their index fingers squeezed musket triggers or yanked cannon flints and shot and fired and shot and fired and shot and fired, before the age of automatic weaponry, creating a rapid and murderous thunder aimed at the advancing redcoats. The British charged the rampart and their guns fired back. In less than two hours, their rumbling lessened and lessened, until just a few bangs rang out. Then, there was silence.

As the smoke cleared, Jackson looked through his spyglass and witnessed thousands of redcoats on the ground, frozen in place where their last breaths were taken. Pakenham lay dead, cut in half by a cannon ball, as Jackson peered over the battlefield with that old bullet inches from his heart. The battle was over, and the least-likely-to-succeed Americans won, barely scathed with only one hundred dead.

The victory was in vain, however, and the British and Americans perished for no reason. What the good Major General Andrew Jackson did not know is that before the Battle at New Orleans started, the war between America and Britain had ended. Two weeks before Jackson and the British fought, a peace treaty had been signed on Christmas Eve in 1814 in the city of Ghent, Belgium, returning the boundaries and policies of both countries back to prewar conditions. But before the invention of Samuel Morse's telegraph nearly two decades later, the message of peace had to travel by boat like a package. The treaty took weeks to reach Washington, not arriving until the middle of February, where it was unanimously ratified on the 16th. By the time Jackson received the official news over two months later, on March 6, the lush, Louisiana greenery had begun to swallow up the slain British redcoats.

On those southern sugar fields, men died needlessly, but the delay of the message had even greater repercussions.

For a brief moment in history, Jackson stood in front of what was best about America—an army of blacks and whites, rich and poor, professional and amateur soldiers, Indians and settlers, and even a few criminals. The differences between these men were numerous, but what they had in common was greater. They all wanted a chance to pursue their happiness, by pushing back the British. During the battle, Jackson promised blacks that they would be paid and respected in a manner equal to whites. On the battlefield, Jackson folded in Native Americans to fight against the British. Near the battlefields, Jackson enlisted women to prepare clothes for soldiers and bandages for the wounded. Jackson unified different people. Out of many, they were one. But this unity and the value that Jackson placed on these souls would not last.

After his victory, Jackson's popularity rose, putting him on the fast track to the presidency. In this capacity, he would remove Native Peoples from their land, exterminating many of them along the Trail of Tears. He would continue the enslavement of African Americans in the nation, amassing great wealth from the slaves on his plantation. And he would ignore the rights of women, by extending voting rights to all white men, and not just property owners. Jackson became known as the people's president, bettering the lives of the people he resembled. Like pieces in a checkers game, all others were sent back, held back, or, in the case of the Native Americans, removed. For a brief moment on the battlefield of New Orleans, Jackson, led and unified blacks, Cajuns, creoles, Indians, and whites. He saved the nation from bondage, becoming a kind of American Moses, but with his broken promises after the war, he turned into an American Pharaoh. Had a message of peace, dispatched with Samuel Morse's telegraph, arrived before this unnecessary battle, Jackson's path to power might have been thwarted, and America might have been different.

Dispatches of Lightning

From the deck of the *Sully,* Samuel F. B. Morse gazed homeward across the Atlantic Ocean, holding back tears. He was sailing back to New York City in a packet ship, which transported mail and goods and moved by the whim of the winds, leaving from the port of Le Havre, France, where the Seine meets the English Channel. Morse's departure was scheduled for October 1, 1832, a date close to his wedding anniversary, which was an occasion that for the last seven years sunk him into sadness. While Morse might declare his stay abroad had been to further his artistic training in painting and to enhance his career in art, his few close friends would whisper that he went to Europe for three years to grieve his wife, Lucretia. Her death in 1825 of a heart attack cut him deeply, and the years afterward deprived him of any peace. He would later confess to his brother that "that wound bleeds afresh daily." What exacerbated Morse's sorrow was that he never got a chance to say goodbye to his beloved; the burden of life after her death became too great, so he escaped across the Atlantic Ocean to Europe. Europe, London in particular, shaped his youth, providing him a time for training in painting and renewing his resolve to be an artist. This time, as a depressed and nearly destitute forty-one-year-old man, Morse sailed to France and then Italy, leaving his three children in the care of relatives and friends, hoping for the healing that accompanies time and space.

Since his college days, when he strolled by the old red brick buildings of Yale, Morse aspired to be a painter. He grew to enjoy what he called the "intellectual branch of the art," filling canvases with murals and historic scenes as the European masters had done. He also expected to live comfortably off the fruits of his labor. Unfortunately, the six-foot-tall stocky frame of his youth was now drawn and wiry. To make matters worse, most of his work was in portrait painting, which Americans loved, but which he viewed as a lesser form of expression. Nevertheless, to earn a living in his twenties and thirties, Morse traveled to many parts of New England, just a few days stagecoach from his parents' home, and to South Carolina, where his mother had family, capturing the likenesses of whomever would pay.

As luck would have it, in January 1825, Morse got the chance of a lifetime to propel his work to the next level within the art world. He was invited to paint a full-length portrait of the famous Marquis de Lafayette, a French commander who fought side-by-side with the colonists during the American Revolution, and who was one of the last surviving heroes of this war. Morse's reverence for Lafayette was second only to that for George Washington, who was a friend of Morse's father, Jedidiah Morse, a militant Protestant minister and also a famous American geographer. Jedidiah never wanted his son to be a painter. Samuel Morse, who was called by his middle name of Finley, never wished to be poor or be so far from home. Even as a small boy, young Morse ran away from his boarding school to be with his strict parents. So when Morse eventually married Lucretia, whom he met while painting portraits in Concord, New Hampshire, he finally had what had been absent in his life, for their bond was an "intensity of attachment." His wife's death paralyzed him and the way he learned about her passing scarred him, too.

In the winter of 1825, during the two long days when Lafayette sat for his portrait in Washington, DC, Morse wrote his wife from his hotel, relaying the festivities he attended at the White House. He closed a letter, dated Thursday, February 10, with "I long to hear from you." Lucretia had given birth to their third child just three weeks earlier and was living with Morse's parents in New Haven, Connecticut. Her recovery was viewed as slow, but seemed sure, and her mood was certainly cheerful. When she retired in the evenings, she spoke about how she looked forward to being with her husband in New York City soon.

A few days after Morse sent a letter to his wife, he unexpectedly received a note that Saturday from his father and knew something was amiss. His father, who was of Puritan stock and never wasted effort on emotions, began with "My affectionately-beloved son." The message continued, "my heart is in pain and deeply sorrow-ful" with the news of the "sudden and unexpected death of your dear and deservedly loved wife." Lucretia never received Morse's letter because the day he put pen to parchment, she had already been dead for three days. Lucretia had an incurable disease, "an affection of the heart," as his father wrote, that ended her life on Monday night. On receiving the tragic news, Morse rushed back to New Haven by stagecoach from Washington. He was in Baltimore by Sunday, reached Philadelphia by Monday night, got to New York City by Tuesday, and arrived in New Haven on Wednesday evening. By the time Morse arrived, Lucretia had been buried for four days.

When Morse returned to his New York City studio to complete the painting of Lafayette, the dark clouds on the canvas might have also illustrated Morse's mood and not just an artistic technique to create contrast with the commander's wide face. Over the next few years Morse went through the motions to establish himself more in the art world, and the fame he gained helped to fill the void. But life overall was flat. While years had passed, for him time ceased to move after Lucretia died. He cursed how slow the news traveled to him. The snail's pace of messages had troubled him before. Many years earlier, when he was a young man in London, he wrote to his parents, "I wish that in an instant I could communicate the information; but three thousand miles are not passed over in an instant." He continued, "We must wait four long weeks before we can hear from each other." The death of his wife only strengthened that yearning for faster correspondences and the potency of his grief forced him to find relief for his sorrow with a trip overseas. With that, Morse made arrangements for his three motherless children to live with relatives and as soon as he could, he headed to Europe.

While on his several-week long journey back home on the *Sully* in 1832, Morse, who grew no closer to the fame and fortune he desired when he was abroad, reluctantly got to know his fellow nineteen passengers on the boat. All the guests met for meals and over the course of the trip became more engaged in each other's work, budding into an isolated community floating in the ocean. One night at dinner, Charles T. Jackson, a Boston physician turned geologist, spoke about science demonstrations on electricity he witnessed while attending lectures in a school of medicine in Paris. He had become deeply interested in electricity and electromagnets, where loops of wire around a horseshoe magnet made it stronger. The young doctor talked at length about an experiment where electricity traveled through wires wrapped many times around the Sorbonne without any loss of time. His fellow passengers had incredulous expressions on their faces. So Jackson went on to tell them about the American hero Ben Franklin, who famously flew his kite during a thunderstorm, but who also transmitted sparks through several miles of wire and noticed that the time of touch and the spark from the other end had no observable difference. One person suggested: it would be great if we could send news in this rapid manner. Morse, emerging from his stupor, asked, "Why can't we?"

The conversation carried on, but not for Morse. An idea zapped him like lightning. After dinner, Morse climbed up to the deck of the *Sully*, got comfortable, and pulled out his sketchbook, drawing out his ideas. All night his mind spun in the cold air about the notion of employing electricity to shuttle messages or "intelligence" across wires. In his college days at Yale, Morse took a physics class with Jeremiah Day, a professor of natural philosophy, who made all the students form a circle and hold hands. Professor Day then applied a shock to one of the young men. "It felt as if some person has struck me a slight blow across the arms," Morse wrote home, and the shock went through all of them simultaneously. Morse wondered that if electricity could travel instantaneously, could a message be sent as quickly, too? Such an invention could help a boarding schoolboy glean affection from his parents or enable an American teenager in London to connect with his home or allow a husband to say good-bye to his dying wife.

The next morning Morse sat at the breakfast table, wearing the same clothes from the day before, reeking with the stale salty scent of night. That autumn evening, in the middle of the Atlantic, Morse thought and thought. On his maritime retreat from reality, secluded from civilization, he formulated how to communicate with the world. Morse's idea, of something he called the electromagnetic telegraph, absorbed him, and he asked Charles T. Jackson questions every chance he got, jotting ideas into his memo book, and trying to concoct ways to send dispatches of lightning. The paintings in his quarters remained unfinished as his muse offered something new—an invention. The *Sully*, like Jonah's whale, provided a container in which Morse could pause, ponder, and proceed in a new direction away from his art and toward an invention. By the time the *Sully* moored in the New York City harbor on the 15th of November, Morse's idea crystallized in his mind.

As soon as he docked and met with his two younger brothers, Sidney and Robert, the only thing he could talk about was his idea for an instrument that sent "intelligence" using electricity. Morse had been away for three years trying to advance his art. His absence meant his brothers collected many stories about his children, the family, and the country. But their three years' worth of news was overshadowed by his oceanic epiphany over the last six weeks. Morse boarded the boat an empty receptacle, grieving and disheartened, but the ocean filled him with a new passion. He gained a new lease on life when he disembarked, but his first priority was to get established again and find a way to support himself and his children, who always lived away from him.

By 1835 Morse had taken a position as an art professor at the University of the City of New York (later NYU), which had all the esteem needed to nurture a man of his ego. Within its safe Gothic walls, he gave high-end art another try and attempted for two years to get a sizable commission from the government for a painting in the Capitol's Rotunda, starting in 1834. As America transformed from a revolutionary dream to a working nation, the capital aspired to be on par with other European cities and needed artwork to demonstrate the causes and conditions that created the country. The commission seemed certain for Morse, but soon that great opportunity was lost, disappointing him greatly. To make matters worse, the number of customers wanting portraits was becoming smaller and smaller. Morse carried himself like a prince, but his pockets were empty. He felt betrayed by his ambition. When talking about art he said, "I did not abandon her, she abandoned me."

What steadied Morse was his secret work on the telegraph. After the stroke of inspiration on the *Sully*, Morse arrived home with a zeal that had left him after the death of his wife. Dutifully, he started putting pieces together into a rudimentary machine to send electrical signals across wires—his electromagnetic telegraph. He called it that to differentiate his design from optical (or visual) telegraphs that existed already, where poles, often located on a Telegraph Hill, signaled messages with their mechanical "arms" in a code called semaphore.

Morse crafted an early electromagnetic telegraph in his New York City artist studio, using the tools around him. He grabbed a wooden frame for stretching canvas for portraits and attached it to a table. He grabbed a pencil and sawed it in half. He grabbed an old clock and removed the inner wheels.

Morse's instrument resembled the equipment in a playground. The frame had a small swing attached to it. Instead of an infant in the swing's seat, there was a pencil. And instead of a mother pushing the swing, there was a horseshoe magnet wrapped in wire, which moved the pencil back and forth, from the pulses of electricity in the wires. From those pulses, an invisible finger pushed the pencil up and down the strip of paper, scribbling in a staccato dance. The resulting trace resembled a third-grader's penmanship sheet on the letter *v*, with characters butted against each other and spaced apart. The bottom point of the *v* symbolized a dot; a stretched line between them meant a dash. These dots and dashes were made by an electrical signal created from the other part of Morse's electromagnetic telegraph—the transmitter.

For sending dots and dashes from the transmitter, Morse borrowed another playground device: a seesaw. As one end of the seesaw went up, the other side, with wires sticking out like a rattlesnake's fangs, dipped down into a pool of liquid mercury, completing the circuit and directing electricity through a copper cable to the receiver. To create the dots and dashes, Morse melted part of his brother's fireplace grate, and poured the liquid lead from it into a mold, accidentally burning part of the carpet, to cast a flat ruler.

He broke the ruler into short chunks with teeth on the edge, like the teeth of a saw, but not all the teeth were there. Where there was a tooth meant a dot; where a tooth was missing meant a dash. From his notes on the *Sully*, Morse created a code of numbers based on the amount of teeth and the space between them. Back in his studio, Morse slid the chunks of metal with these codes under the seesaw. When a tooth pushed up the seesaw's end, it forced the back part down, shooting an electrical pulse to the receiver. The pencil scribbled a *v* onto the strip of paper, advanced by the wheels of the clock. Morse then translated the resulting dots and dashes into numbers and then the numbers into words, listed in a dictionary that he compiled.

These gadgets required much effort to perfect, although they held a paucity of aesthetics, unlike the painter's other works. With the little money he had, Morse bought wire so that he could string up his apartment and have the message travel greater lengths. Once, when he slid the saw-tooth message, pushing the transmitter to complete the circuit, his receiver did not move. He tried several times, tightening parts and making sure that connections were sound. Still nothing.

His amateur knowhow had reached its limit and he needed scientific expertise. In January 1836, he reached out to an NYU colleague, Professor Leonard Gale, an expert in chemistry, who immediately spotted the issue. Just as water in a hose requires lots of pressure to travel long distances, the electricity also needed an additional push to get it to travel far. Morse had one battery, which he learned about from classes with Professor Benjamin Silliman at Yale College, but Gale suggested that he would need several of them lined in a row, like soldiers in formation, each giving the electricity an additional intensity. Gale also examined Morse's horseshoe magnet wrapped in wire. Morse had a few loose loops of copper around the magnet, but what he needed was dozens, if not hundreds, of turns to make the magnets strong enough. Gale read about this in a scientific paper written in 1831 by physicist John Henry, a professor at the College of New Jersey (later Princeton). Morse noted the changes and incorporated them to make it work.

Samuel Morse labored on his art and his invention, but he was also swayed by politics. Growing up staunchly Protestant, Morse detested the pope and the Catholic Church. He wasn't alone. With the huge influx of Irish Catholic immigrants, many Americans feared that their slice of the American pie grew smaller with newcomers. Morse wrote, "Our democratic institutions are suffering." He implored that America should "guard against dangers with which they [democratic institutions] are threatened from the influx of a vicious, ignorant foreign population." Morse's and the mob's anger devolved into a full-blown hatred toward people of a certain national origin and specific religion. He surrounded himself with a group protecting native-born citizens (or nativists), calling themselves the Native Americans. Heartbroken by his life and art, Morse's last love was America, and he had strong ideas about who could and could not be a citizen, even believing that slavery was a divine arrangement of society. To protect his version of the country, Morse ran for mayor of New York City on an anti-immigrant, anti-Catholic platform. He achieved his long-sought fame, and the ungovernable id got an eloquent person to speak on its behalf. Morse lost abysmally, and soon this fickle man, whose father always admonished him by saying "it is impossible that you can do two things well at the same time," returned to his invention. But he never stopped pushing his ideas of who was American and who was not.

While working on his telegraph largely out of view, Morse eventually began to privately demonstrate his invention to students of NYU. His displays worked most of the time, but he was starting to worry. Newspapers reported other telegraphs, particularly a visual (or optical) type in France. Pinpointing the original creator of the telegraph was becoming more difficult, so Morse brought his work out into the open and had the press write about his invention, as a way to set the story straight. His efforts worked, but they also backfired. Not long afterward, Dr. Charles T. Jackson, the electricity enthusiast from the *Sully*, read an article about Morse and insisted that he was a coinventor and that articles going forward should say so. Morse, who surrendered all of his lifeblood into the telegraph, didn't agree. Their acrimony for each other grew, threatening legal involvement, as Morse pushed the endurance of the electrical signal.

On September 2, 1837, Morse exhibited his rustic telegraph to a gathering of friends, students, and professors, successfully sending an electrical signal through one-third of a mile of copper. The strands of wire stretched over a long room that served as Professor Gale's lecture hall. One of the audience members was Alfred Vail, a former student now in his thirties, who was a machinist in his father's ironworks in New Jersey. Vail was smitten by what he saw, and also by its potential. He had a nurturing point of view for wretched things, since he had the heart of a minister, a vocation he worked toward, but his health prevented. Vail, with his boyish face, dark hair, and a saint's patience, was searching for his new life's calling. To him, Morse's instrument looked primitive, but Vail knew he had the healing hands to turn this wooden frame into a metal machine with mechanical and electrical parts. The skillful and calm Vail and the motivated-yet-agitated Morse complemented each other, so they collaborated, becoming a Jacksonian era's version of Steve Wozniak and Steve Jobs.

After months of struggling to fashion the electromagnetic telegraph into metal and to get signals to stretch further lengths with the innovation of a *receiving magnet* (or relay), their luck was beginning to change. Congress printed a circular that solicited inventors to share their best ideas and inventions for sending messages long distances. When Morse read the announcement, he could nearly taste the sweetness from victory. Vail and Morse focused their energies on winning this competition.

As the telegraph became more robust, Morse applied for a preliminary patent (called a *caveat*) in September 1837, to secure a foothold within the legal machinery. On the assembly side, Vail began testing at his father's Speedwell Ironworks in Morristown, New Jersey, which offered more space on its wooden floors and also better machining tools than Morse's New York City rooms. On a cold January 6, 1838, Vail successfully conducted tests using two miles of wire, stretched around the walls, inside the old barn.

As the two grew more confident, they began to demonstrate their apparatus publicly, to prepare for the presentation in Washington. First, they sent messages made of sparks through two miles of wire in front of hundreds of Morristown citizens, then they sent a message through ten miles at NYU, and again at the Franklin Institute in Philadelphia. As their telegraph evolved, Morse dropped his bloated method of converting thousands of number codes into words from his dictionary to a lean code of dots and dashes equaling letters and numbers. With successful tests and a new alphabet behind him, Morse was ready to show the government officials in Washington what his invention could do.

In Washington, DC, Morse started demonstrations of his telegraph on February 15, even presenting it in front of President Martin Van Buren on February 21. His message was successfully delivered over ten miles of wire. Politicians, who never lacked words, had none. Other competitors who auditioned their system to the government conveyed "intelligence" using the slow process of semaphore. Morse was able to have ten words transcribed in a minute, the fastest means of communication humankind had ever seen, rendering those other methods virtually speechless. This victory deserved much celebration. But Morse had yet to overcome the challenge of finding a way to pay for the deployment and launch of his new invention. That would literally take an act of Congress with a bill to finance the installation of telegraph lines.

While waiting for the legislators to grind out the grist of his bill, Morse took a long journey back to England and other points in Europe in May 1838 to acquire foreign patents. He had applied for an American patent in April and felt confident he'd get it. With the sweet smell of victory wafting into his nostrils in the United States, Morse headed abroad to Europe and Russia to see what other winnings might be possible.

Morse tried to get a patent in England, but officials would not even hear his case, for there was a British version of the telegraph there already. While Morse toiled in the United States, there were two inventors in England—Charles Wheatstone and William Cooke—who labored on their own version of sending messages down electric lines. Morse attempted to show the uniqueness of his idea, in an effort to carve out space for his patent. In the English system, messages were made visual with deflected compass needles; Morse employed electromagnets to move a pencil across paper. In the English system, the position of five needles signified a letter; Morse translated a simple code based on dots and dashes. In the English system, six wires dispatched signals; Morse required only one. Additionally, Morse's system scribbled down a message, while the other system did not. The difference seemed obvious to him, but his longing to designate himself as the inventor of the telegraph and publicize his efforts in the press in the United States worked against him. Public announcements of inventions nullify the chances of getting a patent in England. Additionally, the officials did not care to sift through the devilish details of his application. For those reasons, the journey to England to secure a patent was a flop. So he moved on to France.

His outcomes in Paris were no better. While his telegraph was patentable there, France had the additional stipulation that an invention must be actively in use within a year. Morse's attempts to install his system were initially promising, but then that possibility perished. Russia provided no consolation for that loss either. With nearly a year abroad, eating away precious time to improve his telegraph, Morse returned to the United States in April 1839 empty-handed.

In 1840, Morse's American patent was issued on June 20, but he remained no closer to getting the telegraph network installed. After the financial crisis of the Panic of 1837, a cloud of depression cast a dark shadow on all new ideas in Congress and in the country, putting them into a state of hibernation. In the meantime, Morse drew his attention away from his invention and toward another run for mayor of New York City in April 1841, and again on a nativist, anti-Catholic platform. Yet again, he lost; and with campaigning behind him, he filled his free time with his fight for his telegraph. There was still very little movement on the bill, however, so the former political candidate traveled to Washington to drum up interest. He met with one bureaucratic delay after another. "I'm still waiting, waiting," he wrote on January 23, 1843. A few days later, he wrote that the suspense was "becoming more and more tantalizing and painful."

When the bill finally came to the floor a month later on February 21, 1843, it was ridiculed. One vocal representative believed Morse's telegraph to be otherworldly. In the 1840s, the use of magnets in quack medicinal cures was up, and the public understanding of the science of magnets was down. When the vote came on February 23, 1843, Morse's telegraph bill passed by only six votes (89 yeas, 83 nays, and 70 abstentions).

Victory was sugary, but short-lived. Morse's achievement ushered him to the next task of being heard on the Senate floor. But time was running out. The Senate's session would be ending soon and hundreds of bills were lined up before his. Devotedly, Morse kept vigil at the Senate and on the last evening, March 3, he sat in the gallery with his chronic stomach ailment and his last "fraction of a dollar" in his pocket keeping him company. If the bill didn't pass, he worried that, like Sisyphus, he would be condemned to push this boulder of legislation up the hill again—that is, if he didn't starve first. Looking at the heap of legislative bills on top of his, the chances of his bill being heard did not look likely. He could not bear to watch eleven years of work crushed under the pile. Before his bill came to the floor, he lifted himself, as if taxed by Earth's gravity, and walked to his hotel and packed.

When he ate breakfast the next morning, he was greeted by Annie Ellsworth, the teenaged daughter of his colleague, Henry Ellsworth, the commissioner of the United States Patent Office. Morse was always pleased to see her, despite the wretchedness that befell him. Annie came to congratulate him. His bill had passed unopposed in the eleventh hour and the president signed it into law. Morse was awarded $30,000 (about $0.9 million today) to build the telegraph line from Washington to Baltimore, a more than forty-mile distance.

Shaking off his cloak of despondency, Morse experienced the odd feeling of joy when he heard this news. Drunk with this new-found happiness, he offered the feminine courier a gift. He promised young Annie Ellsworth that she could forge the first official message on his telegraph. Now he had to make the wire highway to transmit her words composed of lightning.

With the ink on the bill now dry, Morse had the goal of connecting Washington, DC, to Baltimore with his telegraph system. With the money in hand, Morse put together a team of men, including Mr. Vail, who supervised the equipment; Professor Gale, who provided scientific support; and newcomer Professor James Fisher, who oversaw the wire and its installation. Morse kept a tally on the budget and the schedule. The plan was to bury the wire network underground, with wires inside protective lead pipes. But burying the pipe wasn't easy. After some wrangling, Morse found a capable lead manufacturer and also met a young man named Ezra Cornell, who could use his knife-like plow to cut a trench into the earth to hold the pipe. The trenching process moved forward, but it lagged behind Morse's schedule.

There were more hiccups. By December 1843, Morse had to fire Fisher for the faulty wire and leaky pipes, Gale had to quit owing to his poor health, and winter had to show its dominance over any outdoor project. Morse paused the installation until the spring. But during his break, he was ensnared into a scheme by a trickster politician named F. O. J. Smith, who swindled money out of the government project and into his own pockets. Morse's efforts of untangling from Smith's web made the thorny technical challenges of the telegraph almost enjoyable.

In March 1844, the installation of the wire resumed. But this time things were done differently. Wires were installed overhead and testing of the cables occurred more frequently. As the network came closer to coming alive, Morse and his team came up with a scheme to attract the public to their innovation. The Whig Party, the political party rivaling the Democrats, planned to hold its convention in Baltimore, where the vice president would be announced. Both the press and politicians salivated for this information, which customarily took a day or more to receive, depending on their location. Morse wanted to bring that delicious morsel of information to hungry souls in DC, within a fraction of that time: just a few minutes. However, the telegraph line was still many miles short of Baltimore. So, Morse and Vail came up with a solution. On May 1, 1844, the name of the vice-presidential nominee traveled by hand on a train from Baltimore to the telegraph's end point, where Vail then punched out the news to Morse awaiting in DC. The dark nimbus clouds overhead held back their payload, as Morse's invention graduated from toy to tool. When Vail's transmission arrived, the speed of the message drew more interest than the enthusiasm for the Henry Clay/Theodore Frelinghuysen ticket.

With the system completed, the day of demonstration finally arrived on the 24th of May. Mother Nature made way for this auspicious day by clearing the clouds in the sky, removing the ever-present humidity in the capital, and providing a light wind to cool a nervous brow. Morse's plan was that he would tap out dots and dashes, and Vail would return the same code back. Vail waited in Baltimore for Morse's lightning dispatches from the Supreme Court office in Washington, DC.

As Morse promised, he let Annie Ellsworth select the first official message. Annie implored her mother, a religious woman, to come up with a quotation that spoke to the awe and wonder of this invention, but also to the trepidation it aroused. Mrs. Ellsworth selected a Bible passage (from Numbers 23:23) and Annie gave a paper to Morse, who converted the words on paper to pulses of electricity. He punched in:

dot-dash-dash, space,
dot-dot-dot-dot, space,
dot-dash, space,
dash, space.

Through these dots and dashes, accompanied by the remaining short and long pulses, Vail obtained the code in Baltimore and returned the same message back, marking a new era in communication with the message: "What Hath God Wrought."

The telegraph was an engineering wonder, providing information with dispatches of lightning, but soon the telegraph became part of the social fabric of the nation, weaving it together. Morse's marvel would serve the nation, and in just a few decades also instill a new national habit for consuming information. This became particularly evident during the short term of James A. Garfield, the beloved twentieth president of the United States.

The World at the President's Bedside

Only a few minutes separated President James A. Garfield from the beginning of his summer break from the business of the White House. Starting from the Baltimore & Potomac Railroad Station on the morning of Saturday, July 2, 1881, nearly forty years after Morse's first telegraph, he planned to retreat to his farm in Mentor, Ohio. But first, the president needed to attend his twenty-fifth-year reunion at Williams College, where he'd give a speech and receive an honorary degree. Garfield was also delighted because soon he'd see his wife, Lucretia, who had been recovering by the ocean in New Jersey from malaria. In just a few short hours, the train would transport him to her, like a telegraph message, and they'd be together feeling the breezes by the Jersey shore. For Garfield, this day was long overdue and he couldn't escape soon enough, for the sweltering temperatures of the nation's capital steamed him, like a crab from the nearby Chesapeake Bay. When Garfield sprung out of the carriage in front of the train station, he bounded his burly body up the stone steps of the B Street entrance, and glided past the rows of wooden benches in the small and serene front space of the ladies waiting room. As he headed toward the main hall, he heard a firecracker pop and then felt the searing rip in the skin of his right arm. While deliberating whether to fight or flee, that debate came to a halt when a second pop rang out, and an avalanche of pain in his back seized him, causing his body to thud, knees first, to the marble floor.

James A. Blaine, the secretary of state, had accompanied the president to the train station to squeeze in a few more minutes of business during their carriage ride. These men, both bearded and charismatic, walked into the station arm in arm, lost in a conversation about how they would enter the new Garfield presidency into the history books. But as Blaine witnessed his friend topple to the floor, the euphoric bubble of their conversation burst, and this seasoned politician and orator cried out, "My God, he has been murdered." There were many great and important things to accomplish on Garfield's return from his vacation. But all those dreams and ideas fell to the floor, along with Garfield's body.

The next few minutes were the longest. As Garfield looked up, hovering over him were unfamiliar but learned faces. Nearly a dozen doctors came to the president's side, summoned from the station, from the street, and from nearby practices. An unbearable pain consumed him, making his mind at once cloudy and clear. One by one, doctors turned Garfield over to inspect his injury, and each time a pain rushed over his body, as uncleaned fingers and surgical probes entered his wound. When the misery inflicted by these physicians was over, the doctors did their best to reassure him, although uncertain themselves, that he would survive.

Eventually, Garfield was moved to the White House in a horse-drawn ambulance. Each brick of the road jostled the carriage, pulsating his pain, while his life seemed to drain out along with the blood staining his summer gray suit. At first, doctors convinced themselves that Garfield would live, but with a closer examination and more time, they reconsidered their medical opinion. In addition to the unending pain he was experiencing, Garfield had one thing on his mind: his wife. He asked his close army friend, Colonel Almon Rockwell, to send a message to her in Elberon, New Jersey. When Lucretia opened an unexpected telegram that same day, it said, "The President wishes me to say to you that he has been seriously hurt." The note ended with, "He is himself and hopes you will come to him soon. He sends his love to you." The frail Lucretia, located a few hours away from her husband in New Jersey, had one mission: to get to him quickly. Garfield, positioned on the precipice between the living and the dead, had one mission, too: to see the dawn.

James Abram Garfield was noted by all who knew him to be a kind, honest man, of great will and intellect. He grew up poor on an Ohio farm outside of Cleveland, and studied his way out of poverty. His brilliance was beyond measure, but easy to display. Legend has it that he could translate an English passage into Greek with his one hand and into Latin with the other—at the same time. Garfield became a president of a small college, was a general in the Union Army, and was a House representative for the state of Ohio before he took on the highest position in the land as the nation's twentieth president. At age forty-nine, Garfield, with his stout chest and bright blue eyes, was destined to be one of the country's greatest presidents. Like Lincoln, he was forward-thinking about blacks, and, like Kennedy, he was a charming orator with a celebrity presence. But also like these men, Garfield found himself on the devastating end of an assassin's gun.

Charles J. Guiteau, a forty-year-old drifter, shot the president. Slim and slight at 130 pounds, Guiteau wore a dark suit on that summer day and had a brown beard, a jaundiced complexion, and disengaged gray eyes. Guiteau, who by all accounts was unstable, was a prolific underachiever: He failed at law, at selling insurance, at evangelism, and later at starting a newspaper. He had no Midas touch, but there was no convincing him otherwise.

At the train station, Guiteau had a letter in his pocket admitting that he shot the president as a "political necessity," so that another faction of the Republican party, which Guiteau fanatically supported, could be in charge. Originally from Freeport, Illinois, he bounced around from upstate New York to Chicago to Boston to Hoboken, New Jersey, often skipping out on paying his rent at each hop. Guiteau had hopes of getting an office position within one of the thousands of openings within Garfield's new administration. He was seen at the White House on over a dozen occasions, having his eyes set on being a consul general to Paris. While he was dismissed every time, he never got the hint why. Somewhere along the way, Guiteau got the inspiration to remove Garfield. He once wrote, "If the President was out of the way, everything would go better."

Within hours of when bullets were fired in Washington, DC, every New Yorker knew about it. Telegraph and newspaper offices in 1881 posted messages from telegrams onto schoolroom-sized chalkboards outside their doors, alerting city dwellers of the day's happenings. For those on the farm, people gathered around railroad stations, which had telegraph lines running parallel to their rails. Society was growing accustomed to reading articles from other parts of the country in newspapers. By 1861, when Lincoln was president, news services like the Associated Press transmitted dispatches along tens of thousands of miles of telegraph wires owned by Western Union that crisscrossed the nation. Since the Civil War, communications about battles and other faraway stories were commonplace, traveling along a web of wire from New York to Chicago to Cincinnati to St. Louis to New Orleans to California and all points in between. Newspapers pumped out stories, and a thirsty public drank them in.

When a *New York Times* headline stated "President Garfield Shot by an Assassin," the nation was riveted because people revered Garfield. Although his presidency was only four months old, he had been a much beloved and popular orator since his days in Congress. As he fought for his life, blacks prayed for him, for Garfield believed in the equality of freed slaves. Immigrants on the East Coast prayed for him, for he came up from nothing. The West prayed for him, for he was the son of pioneers, growing up in the rustic Western Reserve. Surprisingly, the South prayed for him, too. Although Garfield was an abolitionist, and targeted their livelihood, he believed in education and in enterprise. The news transmitted by telegraph about Garfield bonded these different groups together.

The next day, crowds stood at telegraph offices, several bodies deep, and were all relieved to see the message on the chalkboard, saying, "a more hopeful feeling prevails." The report further stated, "his temperature and respiration are now normal." Garfield made it through the night. His spirits were buoyed when his wife arrived that evening, pushing the limits of locomotives to get to him. Lucretia, never left his bedside and the crowds at the bulletin boards, growing in number, kept vigil, too.

From the White House, Garfield's devoted private secretary, twenty-three-year-old Joseph Stanley-Brown, had the unenviable job of sending telegraphed bulletins to the press, linking the nation to its leader. Daily bulletins were issued three times a day—in the morning, at noon, and in the evening—giving the status reports on the president's condition. No detail was too dry or too dull. There were reports on how well Garfield slept, what he ate, and what his mood was. For the medically inclined, his temperature, pulse, and respiration were always reported. Most bulletins were short, alerting citizens there was no appreciable change since the last bulletin, or that his condition was favorable.

For the next few weeks, good news prevailed. From the bulletins, crowds learned that President Garfield was cheerful (July 7, 1881); had eaten "solid food" (July 17); was "comfortable and cheerful" (July 29); and had a pleasant nap (July 31). When Garfield had an operation near the bullet hole, doctors informed the public on July 24. These physicians believed that the bullet was a major source of Garfield's trouble and were hell-bent on finding it. They even went so far as to solicit the help of Alexander Graham Bell, the inventor of the telephone, who had also created a metal detector, which made a noise when metal was near it. Bell went to Garfield's bedside in the White House on July 26, listening for the murmurs of where the lead slept. But the assassin's bullet in the president's torso could not be found.

News in the form of official bulletins continued to flow from the White House. On August 1, nearly a month after the attack, Garfield was "feeling better." The president seemed to be recovering, and the nation's hope swelled. There were a few weeks in early August when bulletins stated repeatedly that Garfield had "an excellent day" and on one of them even mentioning Garfield "slept sweetly." The president was surprised by the nation's response, retorting, "I should think the people would be tired of having me dished up to them in this way." But it was quite the opposite; the nation desired to know and also to be in communication with its leader. From the day he was shot, "telegrams from all parts of the country and Europe kept pouring in at the White House," reported the *New York Times*. After the Civil War, America was fractured, but updates sent along telegraph wires about Garfield fused the nation back together.

August in DC in 1881 was hot, and as the temperatures rose, the country's widespread concern for its leader in the oppressive heat rose, too. One bulletin on the morning of August 25 spoke directly to Americans, stating, "the subject of removal of the President from Washington at the present time was earnestly considered." Garfield's doctors wanted to get him out of the oppressive heat and they also wanted to allay the public's concern, but the president was too ill to move. Fever was always nearby, his face was now swollen with an infection of his salivary glands, and he had continual "gastric distress." Garfield, who was a general during the Civil War, told his wife, "This fighting of disease is infinitely more horrible than battle."

Garfield's bulletin reports were mostly optimistic, but his actual prognosis was not. It was believed that the doctors wrote positive reports in the bulletins because Garfield asked to hear them and they did not want to alarm him. The president looked at his charts and stated, "I have always had a keen appreciation of well-defined details and definite facts." He studied his own case as if an outsider. But the words on the charts and in the bulletins and newspapers did a disservice, for a picture made it abundantly clear of his imminent demise. Garfield, who was known for his girth at 220 pounds, withered to nearly half his heft at 130.

In early September, Garfield wanted to be moved to the New Jersey shore to be closer to the sea. Since he was a boy, he'd wanted to be a sailor, but his landlocked home state of Ohio didn't offer much except work on the canals. Crowds lined the path of the train and bulletins kept the public apprised, telling readers that he was eating well (September 11, 1881) and that his cough was less (September 12). By the 16th, his pulse fluctuated in the night. By the 18th, he had "severe chills" lasting an hour, perspiration, and was "quite weak."

In the evening of the next day, and without any warning, a bulletin on the 19th at 11:30 p.m. stated, "The President died at 10:35 p.m." Only a few weeks shy of his fiftieth birthday, President James A. Garfield was dead, after battling infection from his wound for eighty days. The crowds surely wanted to know how he died, and the bulletin provided an answer with "severe pain above the region of his heart." As his body lay in a bed in an oceanside town facing the sea he loved, the telegraph allowed the world to be at the president's bedside.

Garfield didn't live long as a president, but his influence on history, as he lay dying, was profound. His bravery was witnessed by millions of Americans, broadcast live along telegraph cables, making him a reality star of the Gilded Age. The *New York Evening Mail* said, "Lying patiently on a bed of suffering, he has conquered the whole civilized world." Garfield knew his time left was short in September and on a still and contemplative night asked his close friend, Colonel Rockwell, "Do you think my name will have a place in human history?" Rockwell answered, "Yes," reassuring Garfield that he'd live in "human hearts." Garfield would indeed have an impact, but unlike what either of these men expected. The president was the people's patient. With his death, the nation grew accustomed to the frequency, quality, and rapidity of news.

Guiteau would claim that he shot Garfield, but he would also say, "General Garfield died of malpractice." There is a kernel of truth in these words. The bullet in Garfield's back missed his spine, key arteries, and vital organs and was safely tucked in the fatty tissue next to his pancreas. But the scores of unsanitized fingers and unwashed surgical probes in and around his wound carried harmful germs for infections. President Garfield could have been saved had the antiseptic of carbolic acid, evangelized by British surgeon Joseph Lister, been used. It wasn't just a bullet that killed Garfield, but poor medicine.

Garfield was president for two hundred days, and would be noted in history as being killed by an office seeker. In his presidency, he did not get an opportunity to transform the nation, but on his deathbed, he tied the public together, by calcifying the country's compulsion for the consumption of news. Samuel F. B. Morse, the inventor of the telegraph, predicted that wires across the land carrying information would create *"one neighborhood* of the whole country."* As Garfield succumbed to his wounds, messages syphoned from the telegraph fed a nation, drawing disparate communities together that craved to know about their leader's health. Morse understood the power of rapid communication on a visceral level and the urgency of wanting to know what was happening more frequently from his own life experiences. Before Morse tapped out his official message of "What Hath God Wrought," marking a new era, he eagerly sent many more transmissions that were far less poetic, but that also designated the beginning of an epoch. When experimenting on his telegraph and growing more accustomed to rapid exchanges, a bored Morse often sent messages to Vail asking, "Have you any news?"

Just a few decades from Garfield's demise, the telegraph would touch all parts of life and reach all parts of the nation. The telegraph funneled messages along long lengths of iron and then copper. But soon the telegraph, like a container for a liquid, would begin to shape what it contained.

B Brief

Young Ernest Hemingway was a clean-shaven and strapping Midwestern teen who had ambition, but not the kind that went to college. Born nearly seventy years after the telegraph, in 1899, from an early age his mother noticed he was "'fraid of nothing," so after he graduated from high school in 1917, he left the quiet world of Oak Park, Illinois, with its understood rhythm of birth, school, marriage, children, work, and death, and traveled southwest, heading five hundred miles from all he knew. With his ticket and trunk, enthusiasm and unbounded energy, the tall Hemingway boarded a train and arrived on October 15 at the brand-new Union Station of Kansas City, Missouri. For many travelers, this railroad hub was a starting point. But for Ernest Hemingway, it was his last stop. For a brief several months, he came to work at one of the best papers in the nation, the *Kansas City Star*, which unknowingly set him on a course to change the use of American language with the help of the telegraph.

As a cub reporter in a rough-and-tumble metropolis, Hemingway witnessed more life in a few months than during his eighteen years at home. Kansas City was a dish composed of heaping amounts of crime and large helpings of corruption, with a splash of jazz. The muchness of it all strained everyone within its city limits, particularly Hemingway, since he visited the underbelly routinely. As the lowest on the news gathering food chain, Hemingway held interviews at police stations, at crime scenes, and in emergency rooms. His sources consisted of professionals of all types, including doctors, gamblers, policemen, prostitutes, morticians, and thieves. When working on a story, Hemingway dashed to a typewriter in the newsroom and punched out words quickly, before the page was "snatched out" by the copy boy's fingers.

Years later, when reminiscing about his time at the *Star*, Hemingway recalled that the newsroom was a place where he honed his craft. There, he gained, as he said, "the best rules I've ever learned in the business of writing." Hemingway found a mentor, the famous Kansas City newspaperman Lionel Moise, who told him, "pure objective writing is the only form of storytelling." The other advice Hemingway received was not from a person but from a list on newsprint called the *Star Copy Style* sheet, which offered over one hundred tips on how to write. From the outset, this list set the tone for what the newspaper editors wanted. The first suggestion stated:

> "Use short sentences. Use short first paragraphs. Use vigorous
> English. Be positive, not negative."

With its own concise advice, this sheet emulated what it wanted from reporters. Within its three columns, more specific rules stated:

> "Eliminate *every* superfluous word."
> "Avoid the use of adjectives."
> "Watch out for trite phrases."

The news editors hungered for lean language, and Hemingway served up skinny sentences. Newspapers, like the *Kansas City Star*, required economic prose, because the flow of information was limited by the technology at the paper. In addition to the typewriter and lithographic press, prose was shortened by the telegraph.

From the early use of the telegraph in 1832, decades before Hemingway's days at the *Star*, Samuel F. B. Morse frequently scolded his young assistant Alfred Vail as they prepared to demonstrate the telegraph in DC before heads of state. "Condense your language more," said Morse, "leave out 'the' whenever you can." Morse and Vail wrote out their messages by hand and then translated them into dots and dashes before tapping out the messages to each other. For swifter messages, Morse believed Vail must begin with briefer prose and required Vail to trim the fat—removing any unnecessary words that didn't add to the meaning, like prepositions or flowery language. With his telegraph, Morse became the potter of American English.

Morse's telegraph would later profoundly impact the distribution of news. Before his invention, newspapers from different cities sent reporters to boat docks to obtain accounts that crossed the ocean. These correspondents waited for ships, collected the news, and then passed their reports on to headquarters by horse, train, boat, or pigeon. With the development of the telegraph, however, information from faraway places, which required hours to receive, took only minutes. Unfortunately, although this new technology had the advantage of speedy communication, it also had a major flaw. Before clever Thomas Edison enabled two, and later four, messages to travel down the line, telegraph wires could send only one message at a time. So, when a story broke or a news ship arrived, eager reporters rushed to the telegrapher's office. A correspondent from a Boston paper, and a New York paper, and a Missouri paper, and a Virginia paper had to wait their turn, as at a single checkout counter, for their story to be tapped out and sent. To mitigate this logjam in the flow of information, rules were put in place: One rule restricted the amount of time for their transmission (often to fifteen minutes). The other rule stipulated that messages had to be brief.

When telegraph companies formed, they created a pricing structure, which motivated a customer to keep messages short, and to keep the telegraph lines free. They charged a flat rate for the first ten words, and each additional word was one-tenth that fee. With such a pricing structure, a person could send a ten-word message for ten cents (or $3 today) from Washington, DC, to Baltimore. That rate increased with longer distances, so that same message sent from Washington, DC, to Philadelphia cost 30 cents (or $9) or 50 cents ($15) to New York City. Such schemes convinced the customer to condense their communication, and the public got the message. By 1903, half of the transmissions were ten words or fewer and the average message was twelve words. Years earlier, in 1844, when Morse sent his four-word official message, "What Hath God

Wrought," it seems he not only heralded the age with a prophetic Biblical reference, he set a benchmark for brevity.

Businesses employed telegraphs the most, since they could afford such services. By 1887, nearly 90 percent of revenue from telegrams came from commerce (from business correspondences to stock market trades to race track bets); the remainder was newspapers, since personal use was a very small fraction. Only 2 percent of the entire population utilized telegraphs socially for family matters. While business embraced the telegram, society at large dodged it. The cost of sending a telegraph was nearly a tenth of a laborer's weekly pay, so, unless there was an emergency, letter writing was the preferred way to keep in touch. For this reason, when a telegram came to their residence, a feeling of dread came along with it, for it often contained bad news. A sibling living far from home might get a telegram saying, "Father died, come home." Long exposition was squeezed out, like an orange, depulping the feelings and sentiments at the cost of speed. On such a dreadful occasion, more words would have been preferred by the bereaved. But in the calculus of compassion over concision, feelings did not win. The humanness of the message was wrung out to support the telegraph's inability to convey more.

After all, the development of Morse code hinged on the premise of being brief. Morse chose an assortment of dots and dashes for each alphabet letter based on its frequency. He counted the number of letters in a newspaper article and noticed that the letter *e* was the most common, so he assigned it a dot. The letter *i* was the second most popular, so he assigned it two. Brevity also crept into correspondences between Morse and his assistant Vail, who were both of the tradition of writing long, handwritten letters. With this new capacity to rapidly communicate, however, they grew more impatient with deciphering codes for words that did not add to the meaning of the sentence. Their letters to each other, as well as their telegraph messages, became more terse, and a shorthand evolved between them. Morse often sent messages with *t* for "the," *un* for "understand" and *b* for "be." Morse wrote to Vail, "Condense your information, but not so as to be obscure," and then went on to create a nearly indecipherable code, where *i i* meant "yes," *1* meant "wait a moment," and *73* meant "best regards."

Some time later, standard codes were developed for telegraph offices to increase the swiftness of communication. A dictionary of thousands of words, called "The Secret Corresponding Vocabulary," listed a letter prefix and a number to each form of a word, such that w. 879 meant "wire," and w. 889 meant "wisdom," and w. 899 meant "wishful." In 1879, another code used by newspapers soon developed, the Phillips Telegraphic Code, compiled by Walter P. Phillips, a journalist, telegrapher, and later head of the news service United Press. This code became so popular in newsrooms that many abbreviations remain within the American lexicon today. Words like POTUS, SCOTUS, and OK were carved by the telegraph, harkening back to a time where brevity was a virtue.

The constraint of the telegraph chiseled the language of the newspaper and Hemingway, who loved this lean and unadorned prose, embraced it as his own. A little more than six months after starting his job at the *Kansas City Star*, Hemingway left. The Great War was underway and he was itching to be part of the action. He attempted to enlist, but his bad eyesight was undesirable, so he went to Italy to be a Red Cross ambulance driver and took the lessons about writing from the *Star* with him. With time, the success of his books propelled Hemingway's short declarative sentences to a quintessential American style. Generations later, students would soon be encouraged by English and literature teachers to adopt Hemingway's style, unknowingly furthering the telegraph's reach.

The brevity of American Language also originated from a larger thrust within the United States to individuate itself from England.

After the Revolutionary War drew a line in the sand, America also separated itself through language. While both these countries use the same mother tongue, there was a difference in spelling (tire vs. tyre, center vs. centre, color vs. colour). There was also a difference in idioms. For good luck, the British "touch wood," while Americans "knock on wood." There was a difference in pronunciation, too, with British shed-yule for an American sked-yule, a British prih-vah-cee to American pry-vah-cee, and the British vahz to the American vayz. (And, let's not forget the difference in the pronunciation of aluminum, where a Union Jack accent will generate Old Glory chuckles.) Nevertheless, there are intentional and distinct differences in the English language spoken on both sides of the pond. While English spoken by the English consists of loquacious phrases with melodious intonations, American English finds the shortest way to say it. While the British sound erudite and educated, Americans sound agreeable and easygoing.

Just four years after Morse's first message from Baltimore to Washington, DC, in 1848, an anonymous writer in the *Democratic Review* reflected on the influence of the telegraph on literature, and hoped that the telegraph would make the language of the time tighter. "Is it too much to expect that this invention will have an influence on American literature?," they asked. What was apparent was that there was a revolution taking place in the style of the written word, since simpler sentences were replacing complicated ones, which were described as "the sentence within a sentence, armed with all the paraphernalia of comma, semicolon, colon and dash." These sentences were known for "dragging their slow lengths over an entire page," until a period put the sentence—and the reader—out of its misery. This writer hoped that the telegraph was bringing writing closer to perfection with its "terse, condensed, expressive" style. They desired that, with exposure to telegraphic dispatches in newspapers, the public might adopt this telegraphic style with its "Yankee directness." This wish was granted. The telegraph, along with other factors, put an end to those long sentences, much like the long trains that carried Hemingway into Union Station, and replaced them with nimbler vehicles of thought.

As preparations for the demonstration of the telegraph took place in Washington, DC, in May 1844, an irritated Samuel F. B. Morse wanted information more frequently from Alfred Vail. Spoiled by his own invention, Morse grew accustomed to the instant communication made possible by his brainchild. Morse's anxiousness for communication was palpable. When he did not hear from Vail for a few days, he wrote numerous reprimanding letters, saying, "I am somewhat disappointed in not receiving a letter from you." Letters in Morse's era took many days, sometimes weeks, to travel from one point to another, but Morse exhibited an anxiousness to receive them when just a few days passed and he still had not heard from Vail. Today, scholars are troubled by a similar anxiety, as well as an unintended consequence caused by the descendant's of Morse's invention: text messages.

Whenever there is talk about instant communication, there is always a discussion and often a fear expressed about how language is changing for the worse. Interestingly, linguists and scholars today are not so worried about the condensing of language or about the health of the Oxford comma. Studies have shown that students can code switch between writing on their devices and for their assignments. While the shape-shifting of language doesn't trouble those who study language and its structure, many are worried about what this form of communication is doing more broadly. Naomi S. Baron, a linguist and professor at American University warns, "There is something scary happening now."

Samuel F. B. Morse set us on a path for instant communication with his telegraph, which is the ancestor to email, text messages, and all social media. There is a downside to our devices, however. "Online communication," said Baron, "is significantly undermining social interactions." Communication is more than words, cool abbreviations, and cleverly selected gifs. Communication is more than just expressing meaning—communication gives us meaning. And, it is the instant form of communication, where we send messages while apart from one another, that has created a new risk. "It is dangerous because we are forgetting how to be human with each other," said Professor Baron.

When we converse with each other in person, we get cues from one another. When we are online, however, "we are forgetting the importance of taking cues from another person to get a sense of are we being understood," said Baron. By not getting those real-life cues when in the room with someone, we cannot see if they are nervous, or have a blank stare, or want to interrupt us. Gifs, emoticons, and emojis cannot relay this information. Americans, who send more text messages than the rest of the world, are on a fast track to losing something. Without those nonverbal cues, we have convinced ourselves we are communicating well. With those cues, though, we might learn that we aren't. Additionally, being online has created a new nervousness in the twenty-first century, much like the one Morse presented when conversing with Vail in the nineteenth century. "We build up anxieties if we don't respond immediately," said Baron.

A former Facebook executive once said that the website, serving billions every day, was a big mistake. What a Harvard sophomore programmed in his dorm room, creating Morse's "great neighborhood," is actually damaging in lots of ways. What we are losing with instant communication is our ability to read each other's faces and to have conversations. Humans are communal beings. As such, real conversations are better for us than virtual ones. Real friends are better for us than online ones. Face-to-face communication is better for us than the web-based form. Social media makes us unsocial beings. We lose the ability to communicate on other levels besides words. While the Oxford comma doesn't need to worry, society is losing something that is far more important. Communication on our devices—like the original texting machine, the telegraph—has squeezed out the human parts that are not tangible. Fortunately, real conversations add that human part back.

"If you don't interact," said Baron, "you have much less opportunity to develop empathy, and without that in society, what are we becoming?"

4
Capture

How photographic materials captured us in visible and invisible ways.

A Question About a Horse

The request seemed simple enough. A man asked a photographer to take a picture of his galloping horse. While this proposal appeared straightforward, in the 1870s portraits required long sessions at a studio where a person stood or sat steady and smileless for nearly sixty seconds. If that person moved while the lens cap was open, their solid form turned into a blurry ghost in a frame. The limitations of photographic materials made photographers wince when babies came to their studio, knowing that an infant's wriggling transformed them into a dense fog on the lap of a stern-looking mother. If a person made a rapid cameo past a camera, it was guaranteed they'd be rendered invisible. This was why the request for taking a picture of a horse in motion was out of the question. The man making this request wasn't any man, however, but Leland Stanford, a Californian rendition of a Rockefeller, who was unaccustomed to the word "no."

Stanford, after two terms as governor of California, became president of the Central Pacific Railroad, building the eastbound tracks of the transcontinental railroad, twisting politics and profit, and amassing great wealth in ill-begotten ways. Stanford cherished his horses and houses that his fortune afforded him, and it was his home that led him to make this request about a horse. As was customary at the time, Stanford wanted photographs of his palatial Sacramento mansion, and hired forty-two-year-old, Eadweard Muybridge, a San Francisco photographer with a brown Walt Whitman beard, to take them. But soon Stanford's words drifted from interiors of his stately home to the horses in his stable.

Stanford theorized that when a horse ran there was a moment when there were no feet on the ground, in a state called "unsupported transit." But he needed proof. Legend has it that Stanford's millionaire friends mocked this idea, noting that a horse would fall without a foot on the soil, and that this teasing escalated into a hefty bet of $25,000 (nearly half a million dollars today). To save face, Stanford needed a picture and wanted Muybridge to get it for him. While Muybridge might not have been certain that taking a picture of a horse in motion could be done, he was sure that every artist wanted a wealthy patron, he was sure every artist wanted a project that pushed their craft, and he was also sure that every artist wanted a taste of fame. Muybridge had come to the United States from England to make something of himself, changing the spelling of his name (from Edward to Eadweard and Muggeridge to Muygridge to Muybridge) as well as careers before he found the trade where he was unmatched. This request was this photographer's best chance. So, Muybridge put his weathered hand into Stanford's meaty grip, muttering "yes" as they shook, forming a union between two different worlds.

Muybridge was one of San Francisco's best photographers. Fearless, he traveled along desolate parts, like the Pacific Coast, Yosemite, and Alaska, to take pictures. The physicality of his craft suited him fine, as a fit and strong Muybridge carried over a hundred pounds of gear. Everything he needed to take and develop images came along with him. In addition to bottles of chemicals, his equipment included large wood cameras, delicate glass plates, buckets for water, a darkroom tent, expensive lenses, and sturdy tripods. Muybridge, an eccentric man with an unkempt beard and mild blue eyes often appearing in a state of shock, enjoyed being away from civilization with only the company of his wagon mule.

Stanford wished for a picture of one of his fastest horses, Occident, a horse that had captured the nation's imagination with its speed. Occident, and horse racing in general, was a national distraction as a broken nation rebuilt at the end of the Civil War. The country was fractured between north and south, but also between the established east with the newfangled west. Occident glued these pieces together, providing a Cinderella story of a "little horse" that once hauled dirt and became a four-legged prince on the turf, putting California on the map.

In 1872, Muybridge lugged his camera equipment to Stanford's Sacramento stables with the hope of catching Occident in full stride. Muybridge readied the glass plates to make them sensitive to light and able to capture an image inside his dark tent. He poured syrupy collodion onto the surface, fully coating it by rocking the glass plate back and forth, like a chef with a frying pan. Then, he dipped the coated plate into a bath of silver nitrate, which held the picture. Muybridge then put the glass in a holder that was impervious to sunlight and walked over to his camera. Occident trotted a mile, churning forty feet of soil a second, with a driver in a two-wheel cart (or sulky) harnessed to its body. Raising its hoofs in diagonal pairs, it headed to a finish line at a record-breaking speed of two minutes and twenty seconds a mile. To take Occident's picture, Muybridge had to use the glass plates in his camera while they were wet because they lost their ability to hold an image as they dried. The evaporation of chemicals was usually Muybridge's primary concern, but the speed of the horse now surpassed that worry.

When Muybridge took a picture of landscapes, he would set up his camera, with an eager wet glass plate inside, and remove the cap for a few seconds. Then he placed the cap back onto the lens. On his first try to capture the image of a running horse, Muybridge opened and closed the cap as usual. But there was nothing. On the second try, he opened and closed the cap faster. This time the glass had something faint and fuzzy. These results seemed promising. The possibility of finally getting answers about a running horse invigorated Stanford. The chance of adding more newspaper clippings with his name in print to his scrapbook buoyed Muybridge. But the picture was too light to publish and it would take more money and more might to get a discernible one. Stanford opened his wallet, suggesting the use of several cameras in a row to slice motion at different stages. Muybridge complied and was prepared to begin his next set of experiments, until his work was interrupted.

Muybridge had a complicated life. He married a woman and entered a love triangle not of his making. He got out of that love triangle by killing the other man and went to jail, just as he was working on the Occident photo project. Three days later the verdict came back "not guilty," and immediately Muybridge headed to Central America to work on the photography assignment he was hired to do before he shot a man. After several years, in the summer of 1877, Muybridge returned to work on pictures of a horse. He continued his experiments in Sacramento and in San Francisco and then headed to Stanford's Palo Alto Stock Farm.

Before the murder, Muybridge took pictures of a running horse by holding open the lens cap for a moment, which generated a faint and blurry picture. To see a horse clearly and petrify its motion on glass, like an insect in amber, the camera needed to blink faster to capture a shorter nugget of time. To accomplish that, Muybridge created a device out of a cigar box from which he took off two of the wood boards. He nailed these horizontal pieces to two small strips of wood, with a two-inch separation between them, like the rungs of a ladder. These slats were set in a frame. In this arrangement, they were able to slide up and down, like a window, and were held up in place with rubber bands. Then, this whole contraption was set in front of a camera, with the lower slat covering the camera lens, ready to take a picture.

When the horse rode by, Muybridge pulled a string releasing these wooden slats, which fell down like the blade of a guillotine. The opening passed over the lens, in a peek-a-boo fashion, as the second slat came down covering the lens again, allowing light to enter the camera for only a brief moment. With this new shutter, the camera saw a sliver of the motion, and froze the image onto the photographic glass plate. Immediately after the pictures were taken, Muybridge brought them into his darkroom tent to develop them. Above him was a hole in the tent patched with a red rag that transmitted light that wouldn't harm the image.

In his efforts to photograph a horse in motion, he started with one camera, then used a dozen and later two dozen. With fast shutters, Muybridge needed lots of light to come into the camera another way, so that the form of the horse stood out "in better relief" on a glass plate. So he created an outdoor version of a studio on the segment of the track, with cameras on one side and the backdrop on the other. To get more light into the camera, this new backdrop was painted white and angled at a slope, akin to a ladder, bouncing more sunshine the camera's way. The track was also dusted with white powder, making the running surface more reflective and shoving more light into the lens. Muybridge danced with the tools that were around him of silver chemistry, sunshine, and shutters, tweaking them to acquire a snapshot of a running steed. Test after test, the rubber bands soon reached their limit for providing speedy shutters, but fortunately, Muybridge found a solution from the latest technological craze—an electric bell.

In order to capture the horse at smaller fractions of a second, faster shutters were needed. Muybridge turned to electricity to make up for what elastic rubber bands lacked. Electrical appliances were starting to make their way into everyday life, and one invention that made waves in Europe was the electric bell. At the press of a button, electricity looped around an electromagnet, pulling the bell's clapper, causing it to ring. John D. Isaacs, a twenty-seven-year-old engineer who worked for Leland Stanford's railroad company, came up with a way to make a faster trigger based on this new technology. Using the lightning speed of current, an electromagnet yanked a latch holding up a shutter, causing it to fall faster than the blink of an eye. Now, Muybridge's cameras were ready.

Across a track, Muybridge had a dozen wires breast-high above the ground, attached to the dozen cameras. Each camera was spaced apart evenly enough for a complete pace of the horse. When the breast of the horse pushed against a thread, like the winner's tape, the threads were stretched, pulling two pieces of metal together near the camera, and causing electricity to flow into the circuit. This triggered one of the shutters in the shed, creating one image. The horse took a picture of itself, causing the next camera to shoot as another thread stretched. The series of cameras, like an orderly firing range, snapped in a fraction of a second, uncovering the articulation of the horse's body.

Together, the collection of silver-coated glass plates showed each step of the motion. After Muybridge processed the image in the darkroom tent, he emerged with joy, stating, "I've got the picture of the horses jumping from the ground."

There was a faint image on the glass, which he enhanced by hand, producing equine silhouettes on a white background of twelve images of a horse in motion. The picture showed that for a moment a horse runs with four aerial feet. But the endeavor of trapping motion onto photographic glass had a far more lasting impact. Muybridge photographed a fleeting moment and society's hunger for pictures, for capturing every instant, grew into an Everest of pictures.

This all started with a horse.

While Muybridge, an outsider of society, was making headway in the field of photography on the west coast, a pillar of society was also creating photographic innovations on the east coast. Muybridge's name would go down in history, but this east coast inventor never got his due. His name was Reverend Hannibal Goodwin.

Pastoral Blues

Whenever Reverend Hannibal Goodwin spoke, four hundred churchgoers packed the wooden pews in the House of Prayer Church in Newark, New Jersey. Goodwin was a powerful orator who, in the 1880s, resembled a church bell, being both wide and sonorous. This towering, white-bearded preacher with nose-clip spectacles loved his flock and the feeling was mutual. At the end of services, many in his church lingered to speak to him, to be blessed by him, to hear a wise word from him. Over time, however, fewer opportunities arose for fellowship with Father Goodwin after his sermons. As soon as he could, he dashed next door to his home. Goodwin's preoccupation was mostly overlooked by his parishioners, until they noticed his hands, which were splotched with orange-brown blotches that matched the blemishes at the bottom of his white vestments. Soon their whispers swelled to a full-blown gossip about their Episcopalian priest with the regal presence of a prince but the sickly palms of a pauper.

Deaf to the murmurs surrounding him, Goodwin would thrust open the oversized front door of his home and plod up the two flights of wooden steps to his chemistry workshop in the attic. This space was the center of his universe. From 1868 to 1887, he lived ten paces from the church at Plume House, on the corner of Broad and State Streets. On the top floor, in the attic, and below the vaulted ceilings, were ivory-colored walls with stains the same hue as Goodwin's hands. On one side of the room was a fireplace, flanked by two windows. Goodwin cut a five-foot opening in the roof to let in sunlight for his scientific experiments during the day; he toiled under oil lanterns at night.

Whenever his wife, Rebecca, called up to him from downstairs, he responded with silence or short utterances. When he appeared for meals, those were the rare occasions when his wife and their adopted children received his precious attention. He was a man with few passions—food, God, and the religious well-being of the littlest ones of his congregation—and he taught Sunday school in the living room. And it was an incident at Sunday school that caused him to spend all his waking hours in the attic.

Goodwin wanted to show images to go along with the Bible stories he was teaching in Sunday school. So he made an appeal to his congregation and to the diocese to buy a light projection system, called a magic lantern. His prayers were answered and Goodwin was able to get some pictures featuring novel scenes of America's landscapes; however, very few photos depicting scenes from the Bible existed. Fortunately, Goodwin, who was a fan of photography, was more than willing to create his own scriptural portraits on glass slides to show his young flock.

Photography in the 1880s required a person with Goodwin's strong physique. The heavy equipment demanded an elephant's strength yet a spider's grace, as one transported heavy gear, while keeping fragile plates from cracking. When Goodwin found a vista to photograph, he took out a heavy glass plate and dipped it into a bucket full of chemicals sensitive to light, all while inside a dark tent. This prepared the glass for the camera. Sometimes, glass slides were sold already prepared with thick chemicals spread on top, as the art of photography evolved. Once the pictures were taken, Goodwin processed them with other chemicals, to persuade the image to stay. After all that toil and trouble, he had pictures to share with the youngsters eager to hear a Bible lesson in his living room.

Pleased with his creations, Goodwin found that, unfortunately, glass and Sunday school children don't mix. When his helpers inserted these frail scriptural pictures into the magic lantern, often the glass would crack or shatter. After many broken slides, despite the sincerest of apologies, Father Goodwin's patience grew thin and he began to think there had to be a way to create photographs that were robust enough to survive against the most well-meaning of hands.

This quest is what led Goodwin to spend all of his free time in the attic. This quest is what made him a stranger to his family and flock. This quest is what caused the stains on his hands and clothes. He was attempting to create a flexible plastic photographic film that could hold an image and would not succumb to shattering.

A devout Episcopalian pastor may have seemed an unlikely inventor, but he always had a clever mind and a knack for tinkering. Born in April 1823 in the small town of Ulysses, New York, located ten miles north of Ithaca, Hannibal Goodwin grew up on the shores of the Finger Lakes. Goodwin was raised on a farm and was an incurable prankster. As the story goes, during one of Hannibal's more mischievous exploits, an innocent hike with his father led them to being chased by a black bear. Hannibal's practical jokes were never malicious, but merely the preoccupation of a creative mind with not enough to do.

Goodwin's efforts to find his calling emulated the path of a pinball. He entered Yale's Law School in 1844, then Wesleyan University in Middletown, Connecticut, before settling at Union College in Schenectady, taking a broad range of liberal arts classes (from English to chemistry). When he finished his bachelor's degree in

1848, he found God, and attended the Union Theological Seminary in New York City to become an Episcopalian preacher. After Goodwin was ordained, he took positions in Pennsylvania and New Jersey and later Napa, California, before he returned to Newark, New Jersey, to settle in as the fifth rector of the House of Prayer Church.

In 1870, Newark, with its population of 105,000 people, was a manufacturing hub and the home of heavyweight industrialists. Thomas Edison was first located in Newark, before he headed to the quiet enclave of Menlo Park. Newark was also the home of John Wesley Hyatt (1837–1920), who manufactured a novel plastic called celluloid, at the Celluloid Company. Celluloid was beginning to be used as a replacement for ivory for making billiard balls, combs, shirt collars, cuffs, buttons, piano keys, and toys. So Goodwin thought this popular substance might be the material to support his Bible images. Hyatt sold it in sheets, rods, and tubes, as well as in a varnish. Hyatt's Celluloid Company, located at 47 Mechanic Street in Newark, was a mile south of Goodwin's home, so Goodwin took a horse drawn carriage (known as a dinkie) to get some.

Hannibal Goodwin picked up his various chemical supplies to make photographic film of Sunday school pictures. He desired the film to be hair thin. To do so, he used science. "In my college course I had acquired some little knowledge of chemistry," said Goodwin. "With this I began experimenting almost in the dark, putting chemicals and organics in altogether new combinations."

Goodwin desired to dissolve a block of nitrocellulose and then coax it to fall out into a thin layer from the solution like snow in a snow globe. His final formula did just that. He combined nitrocellulose into liquid nitrobenzole, which created a thick syrup, diluted the mixture with alcohol and water, and then poured the concoction onto a glass plate to dry. Each of these ingredients contributed to creating thin plastic layers: Nitrobenzole and alcohol acted like a tortoise and a hare: nitrobenzole evaporates slowly, while alcohol evaporates quickly. The combination of liquids allowed the nitro-cellulose to spread across the glass and fall slowly out of solution, covering the surface like a steady snowfall.

After nearly ten years of experimentation, almost blowing up his attic on one occasion, Goodwin filed a patent for the invention of a roll of plastic film for photographic images. He had just reached retirement age for the church and was thinking about his new life and his financial future, since he rarely saved money—instead spending it all on family, the needy, and chemistry experiments. But the pastor was certain that his invention was useful for photographers after reading magazines claiming a need for a film "as light as paper, transparent as glass." Goodwin's invention met those criteria and also allowed photographers to take pictures rapidly, with a long roll of film. He believed that the filing of the patent would pave his way to a huge financial reward. Little did he know that one of the richest men of the time—George Eastman of Eastman Kodak—was thinking the same thing.

In 1887, after twenty years of service, Reverend Hannibal Goodwin was preparing to leave his post at the House of Prayer Church. Goodwin's health had declined, preventing him from his duties. With his clear calendar, Goodwin kept to himself in the attic, perfecting his photographic films. With time, he was able to get strips of plastic that were a length of ten, thirty, and eventually fifty feet. Soon, he wanted to file for a patent.

As the crocuses broke through the last bit of snow from the Blizzard of 1886, Goodwin put the finishing touches on his masterpiece—his patent application. On May 2, 1887, he submitted to the patent office "Photographic Pellicle and Process for Making Same," describing both the invention of pellicle—a thin layer— and how to make it. This preacher's ability to pontificate at the pulpit, however, did not translate well into this legal document. Most applications were no more than fifty pages of onionskin paper containing precise language. Goodwin's application was as thick as a Bible and read like one. As such, his work waited in a patent examiner's inbox.

Goodwin made many trips to the patent office in Washington, DC, to hurry the application, but the patent office, like his films drying on glass, could not be rushed.

Father Goodwin bided his time working on other inventions, and also wrote George Eastman, a camera and film manufacturer, asking for Eastman to coat and sensitize a seventeen-foot long sample, since Goodwin believed that his idea was protected from being copied or stolen. Eastman, intrigued, wrote back to the preacher, asking a myriad of questions about Goodwin's work. Father Goodwin told the tycoon where he got his chemicals, and also sent some materials to Eastman. This correspondence made the time go quickly for Goodwin as he waited to hear from the patent office, bringing him one step closer to a secure financial future.

In the midst of their correspondence, George Eastman submitted his patent to the patent office on April 6, 1889, two years after Goodwin, on a process of flowing a chemical liquid and evaporating it to get a film. Also, one of Eastman's employees, Henry Reichenbach, a chemist, submitted a patent a few days later, on April 9, for the same process. At the patent office, patent examiners found the three patents from Goodwin, Eastman, and Reichenbach were too similar, so a proceeding to uncover the original inventor began. Eastman withdrew his application so that Reichenbach and Goodwin could battle it out. In the proceedings, Goodwin made his case and brought in samples of his film from 1887. With that evidence, the patent office claimed that Goodwin's application was the original, giving him permission to proceed. Goodwin felt victorious, but he did not understand there was more work to do. When the Eastman Company acquiesced its priority on the invention, Goodwin thought "the matter had settled and ended." He was mistaken, however.

Unbeknownst to the good Father Goodwin, he inadvertently entered into a chess match with one of the largest monopolies in the world. While Goodwin relished his triumph, he became deaf to suggestions he got from the patent inspector on how to improve his patent to ensure it would be issued. The patent officer gave suggestions to Reichenbach, too. Reichenbach narrowed his recipe in the patent to a specific amount of nitrocellulose and camphor. Goodwin didn't make any changes.

On December 10, 1898, Reichenbach was awarded a patent. Goodwin's application was rejected.

The reverend tried to reclaim what he believed was rightfully his and made many more visits to the patent office, spending the little money he had, trying to get some clarity on how to secure a patent. He wanted his patent or a good explanation. The patent office didn't have one, but the patent examiner gave Goodwin some suggestions of how to apply again to give him another shot.

The suggestion was for Goodwin to include camphor in his recipe in the patent. But Goodwin used "not an atom of camphor," and did not care for the "spottedness" it created. Goodwin, however, hoping to get his patent, listened to the patent examiner. This proved to be a mistake. Including camphor into the patent opened up a legal Pandora's box. Now Goodwin needed to prove that his invention was not the popular celluloid, which was made of camphor and nitrocellulose; he couldn't get a patent for something that already existed. This setback was getting Goodwin no closer to his pay dirt. As Goodwin scaled a mountain of paperwork, he wrote to a friend, "I have not only grown older, but much poorer."

Goodwin was now on the losing side of the patent battle versus Reichenbach and entered a cycle of revision, rejection, then repeat. In 1892, 1895, and 1897, he submitted new applications, and each time they were denied. By 1896, Goodwin got new legal representation, attorney Charles Pell of Drake and Company, who made an appeal and sent it to the examiners-in-chief in 1897. Miraculously, a reversal of the earlier decisions was reached on July 8, 1898, which paved the way for Goodwin to get a patent. Goodwin's attorneys were able to prove that all of the changes that Goodwin had made to the application fell within the original description of the patent. Also, Goodwin was able to prove that he produced a film when his patent application was submitted, which he showed in the first interference case. George Eastman testified in that first case that he had not been able to obtain a satisfactory film until 1888.

On Wednesday morning, September 14, 1898, an ailing seventy-five-year old Goodwin made the four-mile journey from his home to Drake and Company and gave a congratulatory talk to Pell and his office to celebrate his new patent. Goodwin found he could still deliver a lengthy sermon on a moment's notice. In his speech, he reminded those in the law office of his belief that "though the gods grind their mills almost hopelessly slow, they generally at last grind out a grist."

As soon as Goodwin got his patent, like David in the Bible, he ran toward his Goliath, Eastman, trying to knock down the giant with stones. Eastman had been making films based on Goodwin's formula and was violating Goodwin's patent. Pell and his firm sued the company for infringement, bringing the preacher closer to a financial nest egg for him and Rebecca to enjoy in their final years. Pell and Goodwin drew up a long list of other companies besides Eastman's that were using the roll film invention so that they could plan their attack of infringement cases. When Pell spelled out his strategy and proposition, Goodwin emphatically responded, "Oh, yes, yes, yes."

The plan was to build a plant in Newark called the Goodwin Film and Camera Company to manufacture photography roll film. A crack in the sidewalk changed all of that, however. In the summer of 1900, Goodwin stepped off a trolley car near his new home on Montclair Avenue, tripped, and fell. His big frame, over six feet tall and 240 pounds, hit the pavement hard, and he broke his left leg. He never recovered from the fall. Goodwin caught pneumonia, and by the end of the year, December 31, 1900, he was dead.

Although still "broken in health and spirit," Rebecca, Goodwin's wife carried her husband's cross, helping to form the company when Hannibal was ill, and merging it with a larger firm to form Anthony and Scovil (later Ansco). The new company continued the fight with Eastman Kodak for infringement, which was taken all the way to the district court in 1902. After more delays and appeals, the case was then settled in Goodwin's favor on March 10, 1914, for a sum of $5 million (over $120 million today), split among Goodwin's heirs and the company. Goodwin was not there to receive what he earned, and his wife, Rebecca was too old and infirm to enjoy it. A few months after the decision, Rebecca died, too.

Hannibal Goodwin fought hard to make images for Sunday school come to life, and ended up living out the story of David and Goliath. The preacher would find that photography wasn't an innocent business of making pictures for children. It led to great burden. Just a few decades after Goodwin's legal battle, photography would be in the middle of a battle again, inspired by school children, but on this occasion a cultural one. This time, the lack of photography's innocence would not just be in the commerce of photography, but built right into the chemical formula.

Under Exposed

In the 1960s, African American mothers noticed something wrong in the seemingly innocent school tradition of the class photo. Every year, youngsters tidied up in their Sunday best for their school class picture, which captured a milestone of childhood. But these black mothers saw something when their children brought these treasured images home. After the Supreme Court desegregated schools with the *Brown v. Board of Education* decision in 1954, the color photo of schoolmates sitting elbow-to-elbow didn't capture black and white children equally. These little ones, trying with all their might to stay still in front of the camera, were being adjusted by the film. White children were rendered as they look in every day life, while African American children lost features of their faces and turned into ink blots. The film could not simultaneously capture both dark and light skin, since an undetected bias was swirled into the film's formulation. For decades, this flaw of the film remained out of sight when schools were segregated and black boys and girls and white boys and girls were photographed separately. But with the integration of schools, black mothers witnessed that color film left their black children in the shadows.

In 2015, two London-based photographers, Adam Broomberg and Oliver Chanarin, excavated this old color film to find out why the film could not capture the likeness of children of all races in a school photograph. When these photographers tested the film they found that "the film wasn't calibrated to deal with that kind of range of exposure," said Chanarin. The film was optimized for white skin. The chemicals to dutifully pick up a range of colors had long existed, ever since the Periodic Table of Elements had become a standard item in most chemistry books. But there was a secret partiality in the combination of these elements used for the film's chemistries, favoring one range of color over another. It was this film's hidden history that was the reason faces in a class photo came out so differently.

Taking photographs in the early days was never straightforward. Early landscape portraits were a challenge not only because of their heavy equipment, but also because most of the chemicals and techniques were homemade. Photographer Eadweard Muybridge captured scenic vistas of California with features of sky and mountains using chemicals he coated onto glass to capture black and white images. By covering the top part of his lens, the bright clouds could be exposed for less time, so that they were not washed out behind the mountains. A generation later, the photographing of black and white images became systematized when Fred Archer (1889–1963) and Ansel Adams (1902–1984) created the Zone Method, with its eleven cards having a gradation from white to black, used to determine the right amount of exposure. A composition containing each of these shades, with more centered in the middle, promised a balance and peaceful coexistence of the brightest whites and the darkest blacks in a photo. When photography switched from early black and white to advanced color film, however, this balancing act got complicated, because there was a juggling of contrast (of black and white) as well as pigments (the colors of cyan, magenta, and yellow). Tweaking one thing shifted another. Shifting one thing changed a third. This burden caused color scientists to create a technique that made life simpler for some, but troublesome for others.

Color scientists designed a cheat-sheet, a color balance card that had the standards of colors to be used in print and in television. With it, a small photo from a camera would look the same when it was printed on a billboard or in a magazine or on a cereal box or in a commercial. This color card was a common feature, like an eye chart in a doctor's office, in the studios of artists, designers, photographers, and camera operators. The most popular of these color charts contained a brunette woman with a forced smile and ice-blue eyes, with a collection of colorful pillows behind her. The job of the artist, designer, photographer, or camera operator was to match the image they saw in a photo or screen to the color on the card. Every object was adjusted—color adjusted—to this standard card, including the color of the woman's white skin. And with that, this card, called a "Shirley card" after the model, fixed the knobs for creating tones and, in turn, made it difficult to render darker skin.

With the simple decision of using this standard color card, skin color that was not Shirley's looked wrong, since the film was optimized to match her complexion. This is why Mediterranean skin or Latin American skin or Asian skin, which contain more green or red or yellow than Shirley's skin, turned out looking alien or burnt or sickly in a photograph. Darker skin would have more extreme outcomes, often appearing otherworldly in some cases; in other cases the film blackened the skin fully. "The technology embodied the ideology in which it was born in subtle ways," said photographer Oliver Chanarin. With the Shirley card, not only her face was the standard of beauty, but also her skin color, causing class pictures to provide similitude and dissimilitude for white and black children.

Early daguerreotype studio portraits, unlike Muybridge's landscapes, were better at rendering a person's image because of the controlled conditions for light, the particular chemicals used, and the increased resolution from the large glass plates. This made early photography an equal opportunity means of reproducing one's likeness. In daguerreotype photography, a simple coating of silver iodide was the gateway to capturing one's face in black and white forever. The only catch was that one had to stand still for a long time, allowing the skin's reflected light to change the bonds of silver on the glass. The chemistry was simple and often homegrown, so a picture could be taken by anyone of anyone.

Photography was touted by abolitionist and orator Frederick Douglass (1818–1895) as the great democratic medium, for, unlike a painted portrait, anyone in any station of life could get an image of themselves. "The humblest servant girl, whose income is but a few shillings per week," wrote Douglass, "may now possess a more perfect likeness of herself than noble ladies and even royalty." In the nineteenth century, Douglass held a deep admiration for this new technology, heralding in his speeches that "Daguerre," Douglass wrote, "has converted the planet into a picture gallery." These words of F.D., as Frederick Douglass was called by his friends, would be prophetic about our age of social media, but what he knew in his time was that images mattered.

In his lifetime, Douglass gave hundreds of speeches in the US and in the UK on the plight of slaves, which he knew from experience, and he would speak for hours to large crowds, as was common in an age before television, and people would pack a picnic and listen. He seemed tireless in his efforts, but when he had no speaking engagements, he would frequently go to a photography studio to have his handsome face captured and let his image do the talking for him, since his pictures were sold and distributed. He let his portraits combat the stereotypical images of African Americans in the 1800s.

By the middle of the nineteenth century, Frederick Douglass was the most photographed human being on the planet. There were more pictures of him than Twain or Grant or even Lincoln. Douglass used his pleasant countenance to counter the unpleasant depictions of black Americans. With laws of the land saying that a slave was three-fifths a person, Douglass sought to show humans, black human beings, at their best. He hoped that whites would see his image as their reflection. While his skin was black, his features were European, since he was of mixed birth. A viewer would see his nose was Anglo-Saxon, his stance was regal, and his posture was powerful. Using his image, Douglass hoped to slay the depiction of the savage.

Something happened at the end of the nineteenth century, however, when Douglass first loved photography, as photographic film moved away from kitchen chemistry and toward a commercial product mass produced by manufacturers, as in the later years of Douglass's life. With the standard formulations from companies moving away from the simple chemistry to sophisticated formulas, the film focused on one set of subjects that could be best captured, while overlooking others.

In the early twentieth century, W. E. B. Du Bois (1868–1963), a preeminent scholar of African American history, also saw the promise of portraying positive images of blacks. Du Bois, born fifty years after Douglass, noticed in his time that, unlike Douglass's experiences, the taking of black images was challenging. He wrote that white photographers made "a horrible botch of portraying them." In Du Bois's day, homemade photographs that rendered everyone equally became less of the norm. The nation had long been smitten by picture-taking, and companies like Eastman's satisfied this need with miles of film as well as film processing services. Film as a consumer product rendered images of one hue of consumer better than another.

Douglass and Du Bois wanted to use images to counter the black stereotypes found in newspapers and magazines of exaggerated facial features of sambos—caricatures with bright, wide eyes and smiles on dark skin. There was a short reprieve from such degrading images with the dawn of photography in the nineteenth century, since early, rudimentary black-and-white photographs depicted reality clearly. As the chemicals for this film, and later color film, were developed around perfecting the portrayal of white skin, however, black faces became underexposed. The only discernible features on them were white eyes and bright teeth on a featureless black shape, inadvertently becoming the damaging stereotype that Douglass and Du Bois so detested. This negative caricature would rear its ugly countenance again in the later part of the twentieth century in the pictures of school children.

Reports in the twentieth century show that Kodak, the primary producer of color film, was made aware of its film's flaw, but dismissed it. Addressing complaints from black mothers in the 1950s and 1960s might have been prescient, since this was the dawn of the Civil Rights Era. Black was beautiful, but the status quo was more. All that changed, however, when large corporations made a fuss about Kodak's film, which they bought in bulk for advertising. A team of two unlikely businesses—furniture makers and chocolate manufacturers—protested against Kodak's films for discriminating against dark hues.

Both industries needed not only for dark browns to come out, but for the details to be obvious and beautifully displayed. A customer needed to be tantalized by milk chocolate, or semisweet chocolate, or dark chocolate that were differentiated in a photo. Newlyweds needed to be enticed by elm or walnut or oak tables plainly shown for their dream home. Kodak employees worked hard to fix the film, making new film formulations and testing them by taking photos, sometimes gaining weight from all the chocolate they photographed. While the complaints from black mothers could not change Kodak, those from these companies could. By the late 1970s, new—and more inclusive—formulations of color film were in the works, and the new and improved Kodak Gold film was on the market by the following decade.

To advertise this new product, Kodak did not want to bring attention to their initial film's bias, so they announced that the new film had the ability to take a picture of a "dark horse in low light." This romantic description was not a reference to the horse, Occident, that Eadweard Muybridge took a picture of mid-stride in the nineteenth century. This poetic phrase was code to signal that darker human skin could now be registered with this new film. This time Kodak distilled the bias out of their chemical formulation, making it possible that dark woods, dark chocolates, and dark skin were able to be captured.

The film captured images, but also the culture's bias. Taking pictures was often a time of levity and of catching a moment of happiness, as a subject posed before a camera. But posing before a camera wasn't always pleasant; it was sometimes oppressive. While many folks in the United States were unaware, American film companies, and the film technologies they created, were used for darker enterprises abroad. This nefarious behavior came to light thanks to a young film chemist in the 1970s, and her actions would make for a better global profile.

Captured

Looking up from her school desk, Caroline Hunter gazed at her tenth-grade history teacher, Mr. Valder, inhaling his words like air. For weeks, "Mr. V," as his pupils called him, had vied for his students' attention during the school year of 1962, attempting to politically awaken these young folks of New Orleans by challenging them to be more active in the civil rights movement. But his urgings went unanswered because he was an interloper of sorts in this segregated part of town. Among the white nuns and the black men and women who taught at the all-black Catholic high school of Xavier University Prep, Mr. Valder was neither black nor clergy. But when he gave the reading assignment *Cry, the Beloved Country*, about life in South Africa, it resonated with young Caroline. She memorized the book's paragraphs, writing them inside her Algebra II book and reciting them by heart. The story in the novel illuminated the hardships under the policy of apartheid, with its segregation of black people, in a place over eight thousand miles away. But those realities resembled her own life at home in New Orleans in 1962. When riding on a public bus heading to school, a sign told her she had to sit in the back. When visiting a department store seeing a dress she adored, a woman told her she could not purchase it. When stopping by a diner smelling hamburgers on the grill, a waiter told her she could not eat at the counter. The book *Cry, the Beloved Country* awakened her and she held onto her memories of it and the lessons of Mr. Valder for as long as she could, until the concerns of teenage life displaced them.

As one of six children, Caroline was raised by a staunchly Catholic mother who imbued in her the importance of doing the right thing and getting an education. Caroline was smart and outgoing, and she had the ability to utter long sentences without the need for much oxygen. With her wide smile, dark-brown complexion, and short-permed bob, Caroline, barely five feet tall, wasn't fond of most pictures of herself. She attended Xavier University of Louisiana, a Catholic, historically black college in New Orleans, majoring in chemistry. She had no time for extracurricular activities, since she worked in the library to help pay her tuition. So when the job offers came in after graduation, Caroline was pleased to have an escape from where she grew up. She had a choice of working in an oil refinery in Louisiana, a drug company in New Jersey, or a photographic film manufacturer in Massachusetts. She headed as far north as she could, starting in the fall of 1968 as a chemist in the Color Photography Research Laboratory in one of the nation's most beloved companies—the Polaroid Corporation in Cambridge, Massachusetts.

In the 1960s, Polaroid equaled innovation, as Apple would two decades later. Both these companies had enchanting leaders in Steve Jobs and his Polaroid forefather, Edwin Land, who was rumored to churn out patents at a rate second only to Edison. Land and Jobs were wunderkinds, and both were college dropouts—Jobs from Reed College and Land from Harvard. Land never dissuaded his worshiping employees from calling him Dr. Land, although he left Harvard after his freshman year and never graduated. This bashful, pipe-smoking genius built Polaroid from the ground up, first making polarizing plastics, hence the name of the company, which prevented glare in headlights and blocked rays in sunglasses. His next big breakthrough was with instant photography, which was the technology on which Caroline worked.

Caroline mixed together chemicals, making the product that was going to be on everyone's Christmas list—instant color photographs. When a picture was snapped, it ran between two rollers as it exited the camera. A soft, buttery paste, officially called goo, residing at the bottom of the white frame of the photograph, would squirt out onto the light-sensitive film, developing the picture. Caroline's concoctions convinced those images to magically appear in under a minute, as if produced by Aladdin's genie.

One fall afternoon in September 1970, Caroline was going to lunch with her new boyfriend, Ken Williams. Ken, a photographer at Polaroid, was a tall, thin, bearded, older African American man. He was self-trained, and one of Polaroid's best artists, with an intuition for his craft not taught in schools. He knew how to pull and push colors in and out, by warming a film packet under his arm or by cooling it in the snow. Ken made his way to the photography department by a stroke of luck, when word got around at the Polaroid factory in Waltham, Massachusetts, about a camera film assembler, who used to be a janitor, who took beautiful pictures. When one Polaroid executive saw Ken's work, he promoted Ken to Polaroid's headquarters in Cambridge. There Ken's job was to show the beauty of Polaroid's products in the art department. It was also in Cambridge where he met Caroline, and they became a Jack Sprat duo—tall to her short, popular to her guarded, modest to her stylish. Although they differed in age and their amount of education, they complemented each other, like a jazz trumpet and a drum.

The two worked in separate buildings on Main Street in Cambridge, three blocks apart from each other and just down the road and across the tracks from MIT, in a part of town called Kendall Square. Ken worked on the first floor of a tall, futuristic glass building. Caroline worked on the second floor of an old three-story brick edifice at the corner of Osborn Street. On the first floor of her building was Dr. Land's office, which was a historic landmark, since the first two-way "long distance" call was received there by Thomas A. Watson when Alexander Graham Bell telephoned him from a room in Boston. This was a room worthy of Land's mighty brain.

As Caroline headed out to see Ken, her long nose was inundated by a range of smells. From the laboratories wafted chemical odors that were strangely sweet, like that of a gas station. Outside the building, the scents of neighboring factories engulfed her, too. A westerly breeze brought the pleasant and sugary aromas of chocolate, mint, or root beer from the nearby NECCO candy factory. Underlying those delightful scents, however, were stenches from the slaughterhouse and the tire reclamation plant.

Ken's office was a den of photographers where grease pencils, loupes, and metal straightedges littered every flat surface. As the two headed out of the office, Ken put on his jacket, to combat the falling temperatures in New England, as the leaves turned yellow. The banter between the two as they readied to go was ordinary and important. But suddenly their conversation was interrupted when they saw something unusual by the door on the bulletin board.

Pinned to the cork was a mock-up of an identification card. The face in the photo was familiar, but the words were not, reading "Department of the Mines, Republic of South Africa." Ken turned to Caroline and said, "I didn't know that Polaroid was in South Africa." She replied, "All I know is that South Africa is a bad place for black people."

On seeing the words "South Africa," lessons from Mr. Valder's tenth grade history class came back to her like an instant image of Polaroid film, along with the memories of the book that deeply impressed her as a teenager. She knew that South Africa was a dark spot of oppression on the globe and wondered why Polaroid had dealings there. The last time the United States heard about the brutality in South Africa was when the Sharpeville Massacre was televised in 1960, ten years earlier, when the police killed seventy protesters. The brutality still existed in the country, but the news wasn't covering it much. The most recent small headline was just the year before, in 1969, when the United Nations wrote a scathing report on South Africa's apartheid policies, recommending that companies and countries "desist from collaborating with the government of South Africa."

While a picture is said to equal a thousand words, the photo on the bulletin board did not provide enough of them, and only generated more questions. By the end of their lunch, they decided they needed to learn more.

For two weeks, Caroline Hunter and Ken Williams went to the library after work, devouring as much as they could about South Africa. Caroline, with library skills honed during her college work-study days, dug up information in many inches of books and in miles of newspaper microfilm. They found that South Africa was a police state and that a passbook controlled the motion of black South Africans. A passbook was a twenty-page bound document with all the information about its holder: where they live, where they can work, and where they could visit. If a person did not have their passbook, they would be fined exorbitantly or jailed for up to a month, doing hard labor. At the heart of the passbook was a photo made by Polaroid.

Not only did the passbook monitor the movement of fifteen million blacks, but passbook laws controlled the flow of black South Africans in and out of white urban centers, like a faucet, changing with labor needs. When workers were needed in the farms, passbook laws were stiffened to keep them tethered to the fields. When laborers were needed during the war effort, passbook laws were loosened to drive labor into city factories. When men were needed in the diamond mines, passbook laws were tightened again to keep them pinned to the quarry. When blacks were not needed any more, the flow was shut off and they were sent back to their segregated enclaves, or homelands, separating whites and blacks.

In 1966, Polaroid created the ID-2, a photographic system that produced two color photographs for identification cards and official documents in sixty seconds, without the need for a darkroom or chemicals. This system made it increasingly easier to create one photo for a passbook and one for a government file. The ID-2 fit in a suitcase, taking hundreds of pictures an hour. One camera in any of the 350 passbook centers in South Africa, along with thousands of boxes of film, could easily capture the likenesses of fifteen million blacks, allowing the country to have information about the whereabouts of each, before the age of GPS tracking.

After reading all they could about South Africa, Ken met with an executive he knew at headquarters on Thursday, October 1, 1970, to share what he learned. During the weeks of their research, Caroline and Ken's fervor grew to a boil. Management's reaction was lukewarm, however, first insisting that they did not know if they had a presence in South Africa and then saying if they did it was small. They asked Ken to find out more and to have another meeting to discuss it. But Ken had heard enough, since he had proof of Polaroid's activities there. He wanted action. He felt that this matter was too urgent for more discussion. Another meeting was set up with Ken for the next day, but Ken didn't show up. He and Caroline decided they were going to do something about it.

That weekend, on Sunday, October 4, the two went to Caroline's work, signing in at the guard station of Polaroid's Osborn Street laboratory, carrying with them a stack of paper. Before their visit to the lab, they borrowed a typewriter and designed a flyer spelling out Polaroid's activities in South Africa. Then they printed out copies using the mimeograph machine at the activist newspaper, the *Old Mole*, on Brookline Street. With the chugging sound of ba-dum, ba-dum, ba-dum, sheets of paper came out one by one with the sweet smell of ink. That Sunday, they posted their leaflets at Polaroid on bulletin boards and the back of bathroom stall doors. They placed the remaining fliers in the executive parking lot. Afterward, they left, signing out at the guard's desk, to enjoy what remained of their weekend and get ready for work the next day.

On Monday morning, as Ken drove Caroline to work after picking her up from her Brookline apartment, they saw flashing lights in front of her building. The Cambridge and Polaroid Police were waiting for them. The high-alert response was partly because of the heightened state of the nation, with regular Vietnam protests and the shooting at Kent State in May. But also, their leaflets could not be ignored, using a slogan from the Black Panther movement typed in the text, and a last-minute addition scribbled on the header stating: "Polaroid imprisons black people in 60 seconds."

Polaroid executives, like befuddled parents with defiant teenagers, eventually allowed Caroline and Ken to come to work, hoping this tantrum would go away.

Caroline Hunter and Ken Williams, calling themselves the Polaroid Revolutionary Workers Movement (PRWM), in the spirit of the civil rights strategies that Mr. Valder once discussed, officially launched their campaign with the mission of getting Polaroid out of South Africa. For these two black employees, changing the course of a company was like pushing an oil tanker with a rowboat. But Polaroid had a weakness: it was protective of its corporate image, making this a battle for public opinion. Polaroid had created an outward-facing wall; it was up to the PRWM to make it topple like the walls of Jericho.

Polaroid struck back with a memo to all employees the next day, October 6, stating they didn't sell cameras to the South African government, with management insisting they had "no company, no investment, and no employees" in South Africa. This was true, in a way. Polaroid had a distributor in South Africa, Frank and Hirsch (Pty.) Ltd., located in ten cities, serving as Polaroid's proxy since 1959. Polaroid had been in South Africa since 1938, benefiting from the cheap labor in the country, and had another distributor, Polarizer South Africa, selling products for them earlier. Ken and Caroline countered Polaroid's claims with another leaflet.

Polaroid's move was next. But Ken and Caroline, in a surprise attack, as taught by Sun Tzu's *The Art of War*, amplified their efforts from paper leaflets to a political rally the following day, October 7. At noon, on the plaza of Polaroid's headquarters at 549 Technology Square, or Tech Square, more than two hundred onlookers under the linden trees listened to Caroline Hunter, Ken Williams, and Chris Nteta, a black South African who was a Harvard Divinity School student. Earlier that day, Polaroid sent a memo to all employees, changing their tune a bit by saying that only sixty-five ID-2 cameras had been sold in South Africa since 1967, and they were used only for military purposes. Nteta, though, was living proof that Polaroid's products were being used throughout South Africa to create passbooks, telling the crowd that Polaroid's written statements were a "tissue of lies."

While the PRWM was small in number, it was bigger than the sum of its parts, for it used a network of activists and the media to spread the word. Newspapers, news wires, and television news programs were hungry for headlines, and Caroline and Ken served up an appetizing dish.

At the rally, the PRWM presented their demands. Using Polaroid letterhead, addressed to Dr. Land, the PRWM described what they wanted: for Polaroid to get out of South Africa, to denounce apartheid publicly, and to contribute profits earned in South Africa to liberation movements. Admittedly, their motives were far-reaching and their actions were seemingly "rabid," but their efforts were no more rebellious than the Boston Tea Party nearly two hundred years earlier. In a heated two-hour meeting with Polaroid's executives the following day, October 8, Caroline and Ken reiterated their ire, spewing out fact after fact about the misdeeds of their employer with their emotions ever-increasing. The next day Ken was fired.

For a few months, Polaroid and the PRWM engaged in an uneven tennis match, as spectators twisted their heads side to side. A week and a half later, on October 20, Ken went to the state capitol to meet with representative Chester G. Atkins to drop off leaflets, and to report the activities of Cambridge's largest employer. *Pock.* Polaroid responded with a press release to the media the next day, October 21, stating that Polaroid refused to work with South Africa since 1948 and were going to explore how to stop the sale of their film. *Pock.* The PRWM served an ace by launching an international boycott against Polaroid, asking people to stop buying Polaroid cameras and film for the upcoming Christmas season. *Pock.* Polaroid volleyed back with a multimillion-dollar marketing campaign around Thanksgiving explaining what it was doing in South Africa. *Pock.* The PRWM was unable to counter Polaroid's big budget, but they kept churning out leaflets.

In another round, Polaroid approached the net, announcing in the press "An Experiment in South Africa" in January 1971, in which they would increase the salaries of black workers at Frank & Hirsch, and provide educational scholarships to its 155 black employees, based on recommendations from a study they did in South Africa in the fall. *Pock.* The PRWM hit the ball over Polaroid's head, with new leaflets, stating that the study could never get a good read on what citizens of the country really want, since it was illegal for a black person to speak against apartheid, a crime that was punishable by death. *Pock.* Additionally, South African laws did not allow a black worker to have a salary greater than any white person in a company, and the education system was viewed by blacks as an indoctrination of inferiority by the government.

In a last-ditch effort to corral Caroline's conduct, Polaroid suspended her without pay on a rainy winter day in New England, February 10, 1971. Two weeks later, on the 23rd, they fired her, since she would not budge. Although she lost her job with a steady salary of $980 a month, and had to collect unemployment at $69 a week for two years, she used every dime she had for the cause. She would buy sixteen-cent stamps to send newsletters to "right thinking" organizations, such as churches and college groups, informing them how to protest Polaroid. As she looked for work, she and Ken would leaflet "early in the morning and late at night."

The movement gained momentum as more voices were added, and the pressure increased with more bodies. Wherever Polaroid, and Edwin Land, were located, so was the movement. When Land was invited to give a scientific keynote address in the Grand Ballroom of the Hilton Hotel in New York City at 2:00 p.m. on Tuesday, February 2, at the Applied Physics Society's annual meeting, Caroline Hunter and Ken Williams were there. Invited by activist physicists, the two voiced their concerns about his technology before he took the stage. Clearly, they were getting under his skin. "The reason why I'm mad at them," said Edwin Land, who was shaken by their presence, "is because they are interfering with my personal goals." The next day, February 3, Caroline Hunter and Ken Williams gave presentations at the United Nations' Special Committee on Policies of Apartheid. When Edwin Land gave a technical talk on color vision at Harvard on March 8, the Ivy League undergraduates, inspired by the PRWM, would not let him speak until he discussed people of color in South Africa first.

The cerebral Edwin Land, who would rather be in his lab, wasn't fond of company politics, let alone global ones. While he was undoubtedly inspired in the lab and made a whole ecosystem for innovation, his thinking about technology within society was less inspired. "You've got to do little experiments," said Land, when speaking to shareholders in 1971 about Polaroid's new plan for their continued presence in South Africa. A scientist by training, Land went on to say that "the function of the physical sciences is to teach the social sciences how to fail without a sense of guilt." Land would eventually learn that scientists cannot separate their research from the application of their research and that the social sciences and the sciences, in general, work best in tandem, like one hand washing the other.

Seven years from the first posted leaflet, Polaroid withdrew from South Africa. The efforts of Caroline Hunter and Ken Williams ignited this revolution by being a pebble in a giant's shoe. Universities and churches applied additional pressure, by divesting their stock holdings in the country and in the company. The final straw leading to Polaroid's withdrawal happened in 1977, when its cameras and film were found being sold to the South African government in a circuitous way. A receipt discovered by a Frank & Hirsch employee, Indrus Nadoo, showed that unlabeled boxes of film were sent to the South African government and billed through a drugstore in Johannesburg called Muller's Pharmacy. Film came into the country clandestinely from other countries, too. Polaroid's departure started the process of dismantling apartheid, like a flicked domino, and Nelson Mandela would come to the United States to thank the PRWM for preventing the further capture of black South Africans.

Technologies we make are not innocuous and their use is not always for the greater good. Technologies, such as photographic film, also capture the issues and beliefs and values of the times.

The vagaries of camera film were not limited to one film manufacturer, either. Kodak film was unable to capture a range of school children, but Polaroid had a similar challenge with its instant film. Pictures taken with Polaroid's ID-2 camera were found to be too dark, since Polaroid primarily designed its product for middle-class white customers. To make up for the products shortfall, Polaroid added a Face Brightener button (a "boost" button) to the back of the camera. When pushed, this button added more light to the flash. Without the boost button, the details of a person with dark skin were barely perceivable, except for their white teeth and bright eyes. Polaroid added this additional button in order for their product to be profitable in the market of identification cards for a black-majority African nation.

This boost button was also explored by London-based photographers Adam Broomberg and Oliver Chanarin in 2015. "Black skin absorbs 42% more light," said Broomberg. "The button boosts the flash exactly 42%." The absorption of dark colors is experienced every summer when beachgoers keep cool by wearing light colors. Darker colors absorb more heat. The same goes for light, which is why the boost button seemed to be intended to illuminate darker skin for photographs.

This bias built into technology has echoes today. Today, silicon pixels in digital photography are not optimized to register dark skin well. Additionally, some web cameras, following instructions from algorithms, are unable to recognize and follow a dark face, but do so easily for a white one. Even interracial couples, who might struggle with awkward family dynamics at Thanksgiving, struggle with getting a great photo together, too. When lovebirds of light and dark complexions want to take a selfie they will find that one will come out, but the other will be a ghost; or that one will come out and the other will be a shadow. Love may be blind, but technology shouldn't be.

What the makers of film and cameras and other technologies have experienced is a tacit subscription to a belief of a standard. In other words, they have gotten on the escalator of "this is how we do things" without asking why. Scholars would describe this type of bias as one that implicitly and uncritically accepts norms and it pervades the cellphones in our pockets. But it isn't the cameras' fault; they are only doing what the lines of code written by humans tell them to do.

These devices capture the biases that exist in our world and, in turn, speak to whom a culture values. As our technologies become more pervasive in our lives, whom they were built for and optimized for will be an important discussion. The goal is to make sure that, moving forward, technology captures what we really want captured about ourselves.

5
See

How carbon filaments pushed back the darkness to help us see, but also veiled our eyes from viewing the impact of its overabundance.

An Enchanted Summer Evening

As the sun sets and a summer day ends, fireflies appear, tendering their tiny torches. With their citrus-colored flashes of yellow, orange, or lime, they signal to children from the east coast to the Rocky Mountains that the summertime ritual of capturing nature's magic in Mason jars has begun.

These six-legged beacons are the great unifier, flashing in parks, backyards, fields, and backlots, enchanting people of all stripes. Fireflies are an embodiment of nature's magic across the globe. Long ago, people in Japan believed that souls of samurais manifested in their lights, inspiring much poetry and art. Today, in Malaysia, stadium-filling crowds stand in long lines for a chance to see fireflies blink in unison, as they carpet tree trunks on riverbanks. And deep in the Smoky Mountains, tourists by the thousands travel hundreds of miles to watch a waist-high aurora borealis in miniature. Fireflies with their little lanterns have shined their way into the hearts of many. But their number is decreasing, by the unknowing hands of humans who adore them.

The culprits are the bright lights shining overhead.

It wasn't always this way. A few short decades ago, instead of too much light flooding our evening skies, it was too little light that was driving the desire for electrical illumination. The darkness worked perfectly for fireflies for eons, but the people chasing after them with Mason jars were not too fond of the dark, and were eager for another way of life at night. In these bygone days, the summoning of clean and stable electrical illumination was left to the realm of dreamers. One such dreamer—and doer—was Thomas Edison, and the creation of the light bulb is often told as a story reduced to a single flash of inspiration with him. Edison, however, wasn't alone in this pursuit of electric lights—or the first. Others were toiling on this problem for years. In fact, Edison hadn't seriously considered working on artificial illumination. But a visit with a little-known inventor drew him to it, inspiring him and creating a world with little darkness.

By the time Thomas Edison was thirty years old, he had already made the world modern with innovations such as the phonograph, the stock ticker, the telephone transmitter, and a telegraph handling four messages at a time. Edison's voracious appetite for inventing was legendary. He promised "a minor invention every ten days and a big thing every six months," a promise on which he delivered. The real firefly he was chasing, however, was his next bright idea. Many scientists across the globe were in competition for making electric lights, which Edison largely ignored while he pursued other inventions. But a visit to William Wallace's home in Ansonia, Connecticut, changed that.

William Wallace was a thickly bearded man in his fifties, who headed his father's copper and brass manufacturing plant, Wallace

& Sons. Known for being deep in thought and for living plainly, William Wallace didn't care for ostentation or attention. He grew up in Manchester, England, but his parents uprooted their seven children to move to America in 1832, when William was seven, to forge new lives and pursue the industrial opportunities in a region of Connecticut where cities were nicknamed by the metals they produced. The Wallace family eventually settled in Ansonia, Connecticut—the Copper City. Here, William dutifully entered his father's trade and helped the company grow. But in the younger Wallace's heart, he was a man of science, hoping that one day the world would know it.

In September 1878, he unexpectedly received a telegram from the great Thomas Edison, asking to visit to see his newest creation. Edison had heard about Wallace's work on his two-month western expedition to the Wyoming Territory with their mutual friend, George Barker. Barker, a professor of physics at the University of Pennsylvania, invited Edison to join a group to watch the solar eclipse on July 29, 1878. There, Barker prodded the Wizard of Menlo Park to visit Connecticut to see a new electrical invention worthy of his prowess. Wallace and Edison had met a year earlier along with the crowds of admiring visitors to Menlo Park. But this was different. This time the young and cocky Edison was intrigued to meet him.

At home, Wallace spent all of his time on the third floor of his large Victorian house in a private laboratory he created that rivaled the nation's best physics departments of the time. The lab contained telescopes, microscopes, and a static electricity machine. He installed a magic lantern projection system to display travel photos and glass slides. On the wall hung a rare picture of the moon, taken from astronomer Henry Draper's telescope. Wallace also had an autograph from kite-zapping Ben Franklin and the trunk of the father of the electromagnetic telegraph, Samuel F. B. Morse. There were so many science artifacts to share with young Mr. Edison when he came to visit in September.

For years, Wallace worked tirelessly in the wee predawn hours on his inventions before the distraction of work. Loyally, Wallace's son, William O. Wallace, helped him run the company, while William Sr. toiled in the lab. Sometimes, Wallace's wife, Sarah, helped him wrap miles of copper wire coils to make generators or electromagnets. Constantly, Wallace's daughter, Eloise, provided a sounding board for his theories. In another time she would have been a recognized virtuoso in the field of electricity, too, for Eloise was as knowledgeable about the work as her father, and gave scientific tours whenever other inventors came to visit. Together, they all made this moment with Edison possible and made arrangements for this auspicious day.

On Sunday, September 8, 1878, the big day of Edison's visit, Wallace waited. After much anticipation, the mechanical doorbell finally rang. Mr. Edison arrived at the Wallace home on Liberty Street along with their mutual friend, the affable and stout George Barker, who peered down at them through his pince-nez spectacles. Much to Wallace's surprise there were others in their party, too, who all seemed to know each other well. The self-taught Wallace also had to entertain prominent scientists, who included Charles Chandler, a chemistry professor from Columbia University; Dr. Henry Draper, the famous astronomer who took the picture on Wallace's wall; Charles Batchelor, Edison's top assistant; and a reporter from the *New York Sun* newspaper, since the press constantly tracked Edison's every move.

Wallace, who normally was a man of few words, chatted with Edison for hours. They broke away from the pack and discussed their hopes to create new forms of illumination that surpassed gas or oil lamps. Wallace expounded on how for several years he came up with ways to bring light to the public and eagerly showed Edison his creation.

The whole group walked up to the third floor for a demonstration. Standing on a plush rug, they watched Wallace turn on a grumbling and jostling generator. Above their heads on the mansard ceiling was suspended a strange metal bracket holding two carbon plates all encased in a round glass bowl. Two thick wires hung from the glass bowl to the floor. There were flashes and then the frame hissed and sputtered out a blinding glow that flooded the entire room like a searchlight. Wallace made light from electricity with an arc lamp, creating a massive spark like the kind a hand to a doorknob produced after one walked across a carpet.

Up to this point, homes were illuminated by oil or gas lamps and sometimes candles, all of which were dim and dirty, and in the case of candles, smelly, too. Wallace harnessed electricity to continuously flash between two blocks of carbon, making a brighter, cleaner form of light.

When Edison saw it, he, like the proverbial moth to the flame, ran toward it. With a childlike glee across his boyish face, he could not control his delight for what he witnessed. The group appreciated what old Mr. Wallace had accomplished, but only Edison saw the future glowing under the glass globe. Edison, always a bit disheveled, spread out the diagrams on the table and pored over them, making quick calculations in his head about the amount of candlelight this arc light system generated. Edison was captivated.

Wallace's moment had finally come. He was on his way to entering the pantheon of electricity demigods, where Edison was Zeus. All those years of his scientific work being mocked as a hobby would be cast aside once it was affiliated with Edison. All his years of living in a gilded cage and unable to pursue his passion fully were about to be erased. All his years of sacrifice were finally going to pay off.

As Edison looked on, Wallace regaled him with stories about his first arc light, made of two carbon blocks on a wooden frame. The bright light formed when a mini bolt of lightning bridged the small gap between the blocks that were each connected to electricity. One day in 1876, Wallace had a man climb the 206-foot chimney of the Wallace factory to attach his contraption. That evening, the glow radiated so brightly that townspeople far away on Division Street reported they could read a newspaper by its illumination. On another occasion, Wallace strung up a row of several arc lights in his factory, replacing oil lamps, which allowed the company to have two shifts: a daytime shift and another that worked until midnight. The arc light provided illumination "each being equal to 4,000 candles," reported the *New York Sun*.

As an invention, the arc lamp had been created before, but it wasn't taken seriously as a source of light. Around 1802, the famous chemist Humphry Davy, at the Royal Institution in London, found that when current was attached to two carbon rods suspended apart, a bright spark, like lightning, appeared bridging the two tips; he called it an arc. Davy didn't think an arc light would be a reasonable means of illumination, however. It was a good parlor trick suitable for his public science lectures. Arc lights resurfaced in history nearly seventy years later, in 1876. Russian telegraph engineer Paul Jablochkoff created a "candle" of two pieces of carbon with a voltage between them. Jablochkoff quit his job in Moscow and planned to display his invention in Philadelphia at the Centennial Exposition in 1876, but he only got as far as Paris. His Jablochkoff candle became the rage in the City of Lights. Professor George Barker saw a "candle" when he traveled abroad and shared this with his colleague Wallace. Once Wallace heard about this new invention, he thrust himself into perfecting it, creating one of the first arc lamps in the United States—and certainly a first for his visitors that day.

The wonder behind the light relied on Wallace's ability to harness the energy from the neighboring Naugatuck River into electricity to power the lights with a generator he built, called the telemachon. In a time when batteries would not provide enough energy for lights, the transfer from waterpower to electricity was key. With the telemachon, "power may be transmitted from one point to another as though it was a telegraphic message," said the *New*

York Sun. Edison was enamored by what he saw and immediately ordered an electric light system and two generators. Wallace happily complied.

When they left the laboratory, they all sat at a dinner table to celebrate, and Edison got his water goblet and inscribed with a diamond stylus "Thomas A. Edison, September 8, 1878 under the electric lights," to mark this day in history.

As Edison departed, he turned to Wallace and gave him a hearty, almost-congratulatory handshake. Then, Edison uttered something that struck Wallace like lightning. "Wallace, I believe I can beat you making the electric light," said Edison. "I do not think you are working in the right direction." Wallace had not only enticed Edison with illumination made from electricity, he introduced him to a world to conquer independently. Wallace's glimmer of hope for reflected glory quickly faded.

Edison's trip to Ansonia put him on the course to build an electric light. Wallace inspired its creation, but just as with any chemical catalyst, Wallace incited an intense reaction, yet his own situation remained unchanged.

September 8, 1878, was supposed to be the best day of William Wallace's life. It wasn't. Instead, it was the day that his own light went dark.

The Wizard's Bright Idea

Edison scrambled back home from his visit to Ansonia erupting with ideas for making electric lights. When the Pennsylvania Railroad train finally reached the small wooden platform at Menlo Park, Edison dashed up the empty red dirt road of Christie Street for two blocks, skipping his own house (and his family), until he arrived at a storm-cloud-gray, two-story building at the top of the short hill. This narrow clapboard structure was longer than a train car and hummed with activity, day and night. This was Menlo Park labora-tory, and Edison its wizard. He hastened up wooden steps to the second floor, reaching a long room with shelves overflowing with jars of chemicals, and told his army of assistants to immediately stop what they were doing. Improvements on the phonograph had to wait. They needed to act quickly.

The visit to William Wallace's laboratory impressed Edison, but what inspired him more was what he didn't see. "The intense light had not been subdivided so that it could be brought into private houses," Edison said. The arc lights in Connecticut shined too brightly, like the flash of an old film camera, and dimming them was impossible. Edison aimed to dice up the light into smaller amounts. However, doing so required a different approach.

What Edison needed was a material that glowed but didn't disap-pear when heated, like something that acted like a hot poker from a fireplace. For generations, civilizations pushed back the darkness by consuming what made the light: torches burned wood, candles burned wax, and lamps burned fuel. What Edison needed was for an object to be incandescent.

The concept of incandescence was not new for a form of light. Starting from 1838, there were over two dozen inventors from Bel-gium, England, France, Russia, and the United States who began this journey, long before Edison. Most of their illumination offer-ings failed, however. Despite this large fraternity of failed incandes-cent brothers, Edison was not deterred. He believed he could learn from their mistakes.

Jumping right into this new electric light venture, Edison formed a new company, read everything he could on past inventions, hired people with the skills he needed, expanded his laboratory, and even held a press conference. He had an abundance of ideas and wired Wallace to hurry up with the telemachons. Within a week from returning from Ansonia, he told the *New York Sun,* "I have it." But he really didn't. Edison guessed it would take him only a matter of weeks or months to subdivide the light; his ingenuity was matched only by his bravado.

Before visiting Wallace in Connecticut in the fall of 1878, Thomas Edison briefly toyed with the idea of electric lights. He had half-heartedly fiddled around with carbon filaments. When sitting at a table, he carbonized paper (baking it to make it pure carbon), attached it to a circuit, placed it under a jar, and used a hand pump to suck out some of the air. When he turned up the current, carbon glowed red and then died, the light lasting only a few minutes. The filament was a sprinter, not a marathoner, because it chemically combined with the oxygen left in the jar and burned. Edison didn't see a solution for keeping carbon from incinerating, so he quickly moved on to other projects and gave up on incandescent lights. On his return from Ansonia, however, Edison embarked on a new quest to bring electric light to the masses.

At the beginning, he tested various metals that glowed with electricity, but his hunt eventually converged on platinum. Platinum was promising: it didn't burn up like carbon and didn't oxidize. But this new metal had a weakness of its own. When a platinum wire was heated to too high a temperature, it melted like butter before breaking and shutting off the light. For months, Edison attempted to prevent platinum from overheating by diverting some of the electricity away from it with complex circuits. But the platinum didn't change.

Glowing wires in a glass bulb, looking like fireflies in Mason jars, filled Edison's lab. Even with the months of effort, however, the platinum filament still didn't work. Edison could not get platinum to glow brightly, and the reason had to do with the nature of this metal. A filament glows because its atoms hinder the flow of electricity that is running through it, and this hindrance, or resistance, heats up the filament like the wire in a toaster. A material that resists the flow of electricity glows better than a material where electricity passes through easily. Unfortunately, electricity traveled without difficulty through platinum. What Edison really needed was a filament made of a different material. Reluctantly, Edison abandoned platinum.

One day in October 1878, Edison swung back to carbon, the element tucked in the cotton thread he tried, using the material he had initially rejected. Unlike what happened with platinum, electricity had a difficult time running through carbon; the thinner the carbon filament, the harder it was for electricity to travel through it, creating even more light than platinum. From his yearlong trials with platinum filaments, Edison learned a few things about improving their performance, namely the importance of a vacuum. A superior vacuum helped carbon filaments survive without reacting with oxygen as they had in the past.

Using high-quality cotton to make the best carbon filaments possible, Edison began a new set of experiments. In late October 1879, he illuminated several electric lamps at the same time to see which worked best. Some lamps were strong; some showed bright spots; some leaked air; some failed for mysterious reasons. One bulb glowed for a whole hour, which extended to two hours, then three, and then finally forty. All of Menlo Park stayed up watching what proved to be the birth of electric lights.

Soon, no corner of the planet would ever be dark again and that reality changed everything.

Electric lights were the result of an old dance. Inventors saw a problem—darkness—and worked obsessively to find a way to fix it. Their invention solved one problem and helped humankind progress in unquantifiable ways. But lights also swayed life in ways the inventors could not anticipate. In little more than one hundred years, artificial illumination has changed the course of the way we relate to each other and to ourselves. It has also changed the very nature of our physical bodies, and those of other species. The beams of light coming from these bulbs moved us in ways seen and unseen.

The Invisible Hand of Daylight

When going for a checkup, patients might do a double take when doctors ask them a new question in addition to how much they smoke, drink, or exercise. That question is, "Are you getting the right kind of light?" This isn't a visit to a practitioner in hippy Haight-Ashbury or New Agey Sedona. This exchange happens in forward-thinking medical offices right now. Today, a range of illnesses is brought about by the lack of exercise, poor diet, insufficient sleep, widespread pollution, and bad genes. But another culprit exists—the light bulb.

Research shows that animals exposed to artificial light succumb to a range of ailments, including an "increase in cancer, cardiovascular disease, diabetes, and obesity," said Professor Mariana Figueiro, director of the Lighting Research Center at Rensselaer Polytechnic Institute (RPI). Animals are not alone. Experts have uncovered that shift workers—those performing the millions of jobs, from security guards to surgeons, at times other than nine to five—have an increase in the risk of cancer and heart disease. By culling reams of data about maladies and connecting them to where people live, what they do, and who they are, researchers have found an epidemiological smoking gun. Ruling out all the other medical factors, one cause for these afflictions is the bright lights beaming over their heads. The lights disrupt their body clock, or circadian rhythm, bringing about these health issues.

In our modern era, with bright lights, we lost an ancient companion along the way—the dark. As a culture, we, like young children, fear it, and do everything in our power to eradicate it. We have streetlights, doorway lights, night lights, but also closet lights, refrigerator lights, and oven lights. There are lighted pathways, lighted signs, lighted doorbells, as well as lighted sneakers, lighted hubcaps, and even lighted toilet seats. And, should there be a loss of power, lights live within our phones. When it comes to illumination, we are never far from it.

Yet scientists now say we have too much light. Specifically, we are experiencing too much of the wrong kind of light at the wrong part of the day, and these lights affect our health. The reasons go back to our anatomy.

Like most people who sat through uninspiring high-school biology classes, scientists thought all that there was to know about the eye over the last 150 years was known. It was commonly acknowledged that light travels to the back of the eye, to the retina. From there, the retina converts that light information into electrical impulses that are sent to the brain, where the brain assembles those pieces together to create what we know as vision. In 2002, however, our understanding of the function of the eye radically changed with a discovery by David Berson of Brown University.

Berson found there is a special light detector, a unique photoreceptor, in the eye (the retina) that doesn't contribute to vision. This part of the eye acts like Paul Revere. Instead of translating the message "one if by land, two if by sea," this photoreceptor informs the body if it is day or night. Just as Paul Revere warned the patriots in the American Revolution to prepare for ground or naval warfare, this part of the eye alerts the body to prepare for day or night. When this sensor detects light—being most sensitive to sky blue light—a message cascades from the eye to the brain to the rest of the body to let it know it is daytime. Specifically, that message zips down the optic nerve in the back of the eye to the part of the brain's hypothalamus called the suprachiasmatic nucleus (or SCN). The SCN telegraphs a message to a small, pea-sized part of the brain, the pineal gland, to stop secreting melatonin, a chemical that alerts the body that it is nighttime. The throttling off of melatonin completes the chemical Paul Revere message that "morning is coming, morning is coming."

Melatonin is an ancient molecule secreted exclusively at night that communicates to cells in our bodies that it is evening. "It is an old chemical that evolved with us," said Thomas Wehr, scientist emeritus at the National Institute of Mental Health. The body needs such a signal because essentially humans are two different organisms—a daytime one and a nighttime one. As a way to conserve energy, our bodies have a time of being "on" and "off." The state we are in is switched on by the lights around us, with melatonin signaling the mode. During the day, our temperature, metabolism, and the amount of growth hormone in the body increase. In the evening, they all decrease and we log off. With artificial lights, however, our bodies don't enter this necessary rest mode.

Eons ago, before the age of electricity, we lived by sunlight in the day and candlelight at night. As dusk approached, our bodies, while awake, prepared for the evening and began to enter our nighttime mode, as the type of light changed from sunlight to candlelight. As the sun began to set, the amount of melatonin began to rise. Today, however, we are unnaturally bathed in the same type of light all the time and the inundation of artificial lights is putting us in a constant daytime mode. The affects are already perceivable: "Modern humans are taller than their ancestors," said Thomas Wehr. "It is partly related to nutrition and other factors, but it is also related to artificial light."

Before electric lights, human physiology was connected to the seasons. More women got pregnant in the late spring and summer. Our bodies tracked the time of year by following the changing length of the amount of daylight between dusk and dawn. With the longer days of the summer, we produced less melatonin than in the winter, and with less melatonin, there was more growth hormone and more opportunity for growth. Today, however, artificial lights nearly blind our bodies from the time of year. "We've almost wiped out our seasonal variation in conception rates," said Wehr. One artifact of this connection remains. "In vitro fertilization is most successful in the late spring and the beginning of summer," said Wehr, since there are longer days with more sunlight and growth hormone.

For humans, being under Edison's electric lights puts us in a perpetual summertime mode with nearly double the growth hormone in our systems than we have on a winter evening. In this constant growth mode, the whole body swims in growth hormone. Every cell is exposed and will respond to this overstimulation. "If you are continuously bombarded with these summer-level growth hormones, with that is a risk of cancer," said Wehr.

Cancer is the malady of our era and discussions about it are hard, for it comes with many uncertainties. For the most part, many researchers believe that cancer starts in one cell. The mutation of the cell for the most part "is just random, it is just by chance," said Richard Stevens, a cancer epidemiologist at the University of Connecticut. So, what has this got to do with artificial light? Well, Nobel Laureate Aziz Sancar found in later work that the "circadian system affects the processes that we know are involved in the causes of cancer," said Stevens. There is a "circadian connection to how our cells will repair the damage to our DNA." The details of what is happening are incomplete, but this work underscores that our bodies have a growth mode and a repair mode, and we need the healing that comes with the dark.

Many factors contribute to cancer; research in this field is one of the most important pursuits of our age. When it comes to women's health, artificial light is one factor that is often overlooked for breast cancer. According to Stevens, "it has been suggested that the pandemic of breast cancer could be explained by the use of electric lights." More studies are needed to understand what is going on, but one population tells scientists they may be heading in the right direction. "Blind women are at a lower risk for breast cancer," said Stevens. "They cannot perceive light at night." Their physiology isn't swayed by the lights. Many medical reports show that blind women are outliers for breast cancer, but much more work is needed to understand how the glow of artificial lights affects women.

Poets say that the eye is the window to the soul. Scientists would say that the eye is a clock, or rather, like the reset button on a clock. Our bodies contain an intrinsic, built-in rhythm where we anticipate the onset of the day, but our body clocks lag by about twelve minutes. A day elapses in twenty-four hours; our internal clock averages 24.2. If we were placed in a dark cave with no visual cues, we would slip behind the solar day, like slow, antique timepieces. When we see light in the morning, however, specifically its sky blue light, our biological clocks synchronize with Earth again.

The sensitivity of the Paul Revere photoreceptor to the color of sky blue is a clever choice by nature and makes sense biologically. The best way to inform the body that it is daytime is to specially tune part of the eye to this iconic color, just like tuning to a specific radio station. Mother Nature could have used all of white light, which is made up of a rainbow of colors (red, orange, yellow, green, blue, indigo, and violet). A lightning storm, with white electric bolts, however, could have accidentally drawn our ancestors from nighttime mode to daytime mode. Sky blue exists unequivocally in the day; it is a unique signal to tell the body that it is time to switch.

Unfortunately, artificial lights do not fully emulate the light from nature, from the sun. The mighty sun contains all the colors of the rainbow in its light. Artificial lights contain only part of the spectrum; incandescent bulbs are redder, compact fluorescent and LED bulbs are bluer. So how can modern citizens live well under the glow of artificial lights and correct the course charted by Edison? The prescription is simple. According to cancer epidemiologist

Richard Stevens, we need "dim evenings and bright mornings." The day must start with body clock-resetting bright blue light in the morning. "Taking a walk is the best thing. You get exercise and you get a bolus of the bright blue light," said Stevens. For those who are indoors, LED lights and bright compact fluorescent bulbs are pretty strong in the blue range of light, too.

Having lots of blue light during the course of the day is good. But the type of light must change as the day progresses. "Light in the morning will have one effect on the body. If [blue light is] given in the evening or the middle of the night, it will have a negative effect," says RPI's Professor Marianna Figueiro. This is why the color of the light must change later in the day. We need redder light in the evening. This includes reducing the blue light from our computer monitors, television displays, and cellphone screens. "Starting at dusk, dim [the lights] way down, and have a light source that is incandescent," said Stevens.

What might help us mitigate this modern sea of lights may be new technologies. Smart bulbs are on the market that can become redder or bluer with the turn of a dial. Additionally, wearable technologies, like those developed in Mariana Figueiro's laboratory at the Lighting Research Center at Rensselaer, will inform us of what kind of light we need and monitor our "circadian light," with a tracker for illumination. With this system, an app will say "get more blue light, or remove blue, or go outdoors," said Figueiro.

Scientists have some advice for those who wake up in the middle of the night, too. According to Stevens, it is best to "stay in the dark," he said. "It is much easier to get back to sleep." This wisdom was commonplace hundreds of years ago, but no one knew it. When our ancestors woke up in the middle of the night between their sleep segments, they did things around the house by candlelight, such as eat, pray, read, or do housework. What we know now is that even when they were awake, their bodies were still in a nighttime mode. Candlelight is a dim, reddish light that does not trigger the increase in melatonin. When a bright electric light is turned on in the middle of the night, melatonin plummets as soon as you can measure it. "If you turn off the light in five minutes, the whole thing will rebound," said Stevens. "If it is more than 20 minutes, you're cooked."

To improve human health, people need to get the right type of light at the right time of day. This is not a mystical claim, but a medical certitude. "Light is the driver of your biological clock," said Figueiro. "It drives everything in your body." And for that we must not look at light bulbs as innocuous objects glowing in the background, but as prime movers of human health.

Edison ushered in an age of lights, but what society needs is to get back in touch with the dark—for reasons beyond our health. Stars have been humankind's longtime companions, helping mariners and pioneers navigate. For centuries, humanity saw thousands of stars. Today, city dwellers can see about fifty. The reason is that most Americans are exposed to brighter than normal night skies with artificial lights. Within just a few generations, the blackness of the night has changed. When our great-grandparents were young, cloudy, moonless nights were the darkest evenings of the entire month. Today, cloudy, moonless nights are some of the brightest, as the water droplets and the dust within the clouds bounce and reflect light, like a disco ball.

Above us, and without our knowing, is a breathtaking scene. But the sky glow, created from the lights, has obstructed our view of this celestial movie over our heads. "It is like you are at the cinema, and you have the lights on," said astronomer Fabio Falchi. As such we cannot see the film's details because we "lose the contrast of the screen."

The fact that the pristine night sky is foreign to us became all too clear in 1994. When the Northridge earthquake hit Los Angeles on a January morning, there was a complete blackout that evening, removing the glow from the lights. Many tense Los Angelenos saw something strange in the sky and called 9-1-1 to report a "gray, silvery cloud." What these southern Californians were seeing was the Milky Way. Reports say that two out of three people in the United States cannot see this formation any longer.

The night sky we experience today is very different from that of our grandparents or great-grandparents. While we gained something from the ease of illumination, we forfeited something, too. "You lose out on the experience that for all of human history has inspired us," said Paul Bogard, the author of *The End of Night: Searching for Natural Darkness in an Age of Artificial Light.* Most of us never see the night sky in all of its glory because the streetlights place a veil over our eyes. A true night sky is a dizzying three-dimensional experience, with stars of different brightness and colors, like van Gogh's *Starry Night.* When we see the night sky in all of its glory we are "stepping outside our door at night and coming face-to-face with the universe," said Bogard, who visited some of the darkest places on the planet when writing his book.

A feeling of self-importance came along with the installation of more lights. When you come face to face with the universe "you realize, you are really small," said Bogard. Artificial light took that awe away. With the universe now invisible to us, it is easy to incubate hubris under these lights. The dark sky used to be a window. Today, it is a mirror.

The Firefly in the Coalmine

Our friends the fireflies started us on this journey. Fireflies, which are also called lightning bugs, are neither fly nor bug. They are beetles that, besides being a tasty meal for some birds and spiders, serve no indispensable function for all of nature, like the pollination of plants provided by bees or the aeration of the soil by ants. While their role may be limited, fireflies, with numbers of species in the thousands, have cornered the market on wonder. As nature's magic lanterns, they are enchanting not just for their light, which was miraculous before Edison, but, in our modern age, for their ability to pull us away from our distractions.

Fireflies speak with a Morse code of flashes—like summer campers communicating after curfew. They glow by a chemical reaction called bioluminescence. The chemical cocktail of oxygen, a molecular energy packet called ATP, a light-providing luciferin compound, and the luciferase enzyme create a molecular flashlight. Those firefly messages, however, are not innocuous; they are dispatches of love. Hovering knee-high above the grass, the male firefly announces himself, flashing a message, identifying his gender and specific species. While no human is fluent in the firefly language, the best guess is that a firefly may be saying something like, "I am a male and I am a *photonis greeni*," explained Sara Lewis, a professor of biology at Tufts University and the author of *Silent Sparks: The Wondrous World of Fireflies*.

Meanwhile, the female firefly perched below on a blade of grass or shrub leaf, looks up at the male's sparks. If she likes what she sees, she'll coyly respond with flashes loosely translated as, "I like you," as Lewis explained. Once the male gets the "green light" that she is interested, the flying male firefly halts in midair, drops like Wile E. Coyote to her vicinity, and makes the hour-long trek to her blade of grass. When they meet, that's when the fireworks really begin.

This critter courtship relies on the ability to see each other. When artificial lights glow high above, they shine so brightly that the female firefly cannot see the male flashing. A male will blink at her, but, because of the glare, she won't know to flash back and these potential lovers may never meet. Additionally, the lights stiffen the competition. Females prefer male fireflies with very bright lanterns, which show a male firefly to be virile, with good health and good genes. The outside lights in the background, however, make the male's beacon look dimmer than it actually is, rendering the female uninterested. So, she doesn't blink back.

The bright bulbs of the human world are masking the essential mating signals of the fireflies and causing a failure to communicate. Male fireflies can possibly flash brighter to find and attract a mate, but they use up their precious energy to do so. Fireflies in their adult stage have a life of no more than fourteen days. For some species, they spend two years underground as larvae, eating and growing, eating and growing, and storing up energy. The energy they save allows them to glow, where one molecule of ATP produces one photon of light. Adult fireflies live off their energy reserve. They rarely eat, since they have a very short time to see and be seen and find love.

Fireflies are not the only creatures that wish we would turn down the lights. Birds, insects, and sea turtles are among the many, many animals that wish we would, too. What most do not know is that "nearly two-thirds of insects are nocturnal," said Paul Bogard, author of *The End of Night*. All their activities change with artificial lights. For some insects, like moths, being drawn to a flame is not poetic, but a punishment. Moths circle a light source and die of exhaustion from it. Blinking lights on communication towers enchant birds, for reasons still unknown, causing birds to fly around them and suffer the same demise as moths. "About 6.8 million birds a year in the US and Canada die this way," said ecologist and USC professor Travis Longcore. For insects, that number is in the billions. This loss has implications for the entire ecosystem. Insects feed other species higher up in the food chain. A chain is only as strong as its weakest link and our electric lights are undermining that chain for all animals.

Artificial lights can cause sea turtle hatchlings to make a devastating choice. As they emerge from their shells on a beach's shore in the evening, they have a few moments to figure out the direction of the water, which offers shelter from predators and protection from dehydration. Instinctively, they know to head toward where there is the most light. For generations that direction was the water with the moonlight shimmering on its surface. Today, however, the direction that is the brightest is often away from the sea and toward the glow of the city.

Despite this dismal prognosis, fireflies and other wildlife can easily be saved. According to some vocal astronomers and the International Dark Sky Association, all it takes is being mindful about the lights by applying a cover around fixtures so that the light goes downward, by illuminating specific areas at levels that are actually needed, and by using lights on demand with smarter bulbs.

Light can be applied where needed without sacrificing efficiency or design and still provide enough illumination to see. New Yorkers who stroll on the elevated walkway of the High Line will climb stairs, barely noticing that the banisters have two additional functions: They hide the lights illuminating the stairs, and they also guard the light from contributing to sky glow by facing the light downward. Thoughtful creatives are intentionally taking another look at designs to save the night sky. Additionally, some designers are enabling parking lot lights to have a low setting when they are not in use and a high setting that is activated instantly when motion is detected. Lights on desolate streets could follow suit. A small part of the night can be saved, and the cost savings could be tremendous. According to author Paul Bogard, reducing outdoor lighting can have "a savings of 100 billion dollars worldwide."

Although gas stations are ten times brighter than they were twenty years ago, humans do not need very bright lights to see. The eye adjusts to the brightest object, functioning superbly with low lighting. The reason lies in the science of the eye. Within our eyes are rods and cones in the retina, which sense light. Rods are our superior night vision goggles that image the world in black and white. Cones are active in bright light and see the world in full Technicolor. While there are six million cones in the eye, there are 120 million rods to help us make out shapes and images in the dark. Most of us live in a world where we rarely use our rods at night, however, needing only our less sensitive cones instead.

We as a species are deeply afraid of the dark, which has fueled our addiction to bigger, brighter, and bolder lights. But as a result, we are harming the animal kingdom, and we are also harming ourselves.

As we get older, we don't see light in the same way. Scientists have shown that the lenses of our eyes are less able to transmit blue light through them with age. The eyes of a twenty-five-year-old lets almost all the blue light in; only half of the blue light that makes it into a twenty-five-year-old's eye will reach the retina of a person who is sixty-five years old. The remaining light can create glare. "If we put lots of blue light in our streetlights," said Fabio Falchi, an astronomer and an expert on the brightness of the modern night sky, "this is not positive for security if you consider how the population is getting older." As cities explode with the number of LED street lamps, which produce lots of blue light, these lights are actually handicapping senior drivers, by using the part of the light spectrum to which they are least sensitive.

Many will claim that more lights lead to less crime. While anecdotally that seems to ring true, few findings actually support this. In 2008, the energy company of San Francisco, PG&E, found "no link between lighting and crime," as stated in Paul Bogard's *The End of Night*. They found if there were any links they appeared to be "too subtle and complex to have been evident in the data." Some level of light may prevent crime, but eventually there is a tipping point when too much light gives rise to glare, making it more difficult for a potential victim to see an attacker.

We need to be smart about how we use lights, by dimming them, by blocking upward light with a shade, and by using them only where we need them, as recommended by the International Dark Sky Association, but also by removing the blue light they produce, as recommended by the American Medical Association.

Sunlight has all the colors of the rainbow in it, but LEDs are rich in blue. From the point of view of the medical community, LEDs in general are not bad; however, LEDs containing lots of blue light are. As of 2016, 10 percent of city streetlights had been converted to blue-rich LEDs, and the efforts to change to these bulbs has accelerated. Understandably, LEDs are the poster child for a city's cost reduction efforts, since they have greater efficiency, provide more illumination, and have a longer lamp life. Cost saving is an important posture; however, the LED bulbs that are being installed are not the best suited for human health. LED manufacturers have developed bulbs with less blue in them but those are not in city lights today.

To reduce light pollution, it will take the efforts of designers, entrepreneurs, citizens, and cities to make changes and help instill a new national habit, and help society move away from bigger and brighter lights and toward healthier ones. Our enthusiasm for installing LED bulbs, without consideration of their impact, can be likened to developing a new automobile. The focus is on "having an engine that gives you many more miles per gallon," said astronomer Fabio Falchi, "so we increase the efficiency of the engine, at the expense of higher pollution."

The issue is that the impact of lights, as well as our infatuation with lights, is invisible to most, which is why scientists, like astronomer Fabio Falchi and his coworkers, have mapped out the light pollution for everyone to see. Using satellite images, they found that 99 percent of people in the contiguous United States live in light-polluted areas, and that light spreads out. "One can see that Chicago's light extends into the Great Lakes," said Falchi. There are also some surprises on the map. "The sea between Japan and South Korea is one of the brightest spots on Earth," he said, "because the lights are used to attract squid." It might literally take an act of Congress to reduce the blue and dim the lights. But with more cities and states changing their practices, federal legislation is not impossible. It has happened before: in 1978, the chemical lead was removed from paint, since it is a neurotoxin linked to developmental problems in children. Installing lights that are less bright and less blue is possible with a similar consciousness, effort, and education.

With such coordinated thrusts, our futures will be brighter—brighter in the right way.

Figure 1

Ruth Belville at the entrance of the Royal Observatory in Greenwich where she obtained the accurate time before dispatching it throughout London by foot.

Figure 2

Arnold the Clock was the Belville family's pocket watch and was used to distribute time to London customers for over a century.

Figure 3
Benjamin Huntsman's name was a marker of high quality crucible steel, as this advertisement from 1900 shows. (No picture of Huntsman exists.)

Figure 4
A twentieth-century Sheffield workman treads the clay for the crucibles that will hold the molten steel. Treading the clay was a reliable way to detect pebbles and air pockets, which caused cracks and leaks.

Figure 5

A Sheffield worker pours the hot metal from the crucible into a mold. To keep the metal clean, the unwanted particles on the top of the liquid were prevented from falling into the mold.

LONG DISTANCE TELEPHONE CONQUERS TIME AND SPACE

Figure 6

A crowd of New Yorkers gathers around the window display for "The World's Most Accurate Public Clock" to set their watches. This clock kept time using a quartz gem and was built by Bell Labs scientist Warren Marrison.

Figure 7

The clock face of "The World's Most Accurate Public Clock" was nearly three feet in diameter. The second hand was much longer than the minute hand, so that onlookers could precisely synchronize their watches to it.

Figure 8

Warren Marrison sitting beside one of his early quartz clocks—a crystal chronometer used for scientific experiments. Marrison ushered in a new age of timekeeping, yet he is often overlooked in history.

Figure 9

At the heart of Warren Marrison's clock was a ring-shaped quartz gem that vibrated within an electrical circuit to provide the accurate time. This quartz gem is about one inch thick.

Figure 10

At Bell Labs, Marrison's clocks sat on special tables to reduce vibrations from New York City traffic. Some of his early quartz clocks didn't have clock faces, but used counter dials.

Figure 11

Marrison worked in this Manhattan building and created his quartz clock on the seventh floor. This building, located at 463 West Street, was the former home of Bell Laboratories.

Figure 12

The "Lincoln Special" transported Lincoln's body across a nation in mourning. Crowds could identify this train by the portrait of Lincoln on the front and from its half-muffled bell.

Figure 13

Lincoln's presidential rail car was designed to be his Air Force One, but instead it became his hearse.

Figure 14
A large crowd stands in the rain at the Camden Street train station in Baltimore waiting for Lincoln's funeral train to arrive.

Figure 15
Sir Henry Bessemer was a British inventor who created a steel-making process by removing the excess carbon in cast iron with a blast of air.

Figure 16
William Kelly was an American inventor who blasted air onto molten iron to reduce the cost of fuel—a method he called a pneumatic process.

Figure 17
A Bessemer converter was used to create steel with a blast of air.

Figure 18

A map showing how far and how fast one could travel in 1800. (After *Atlas of Historical Geography of the United States*, used with permission)

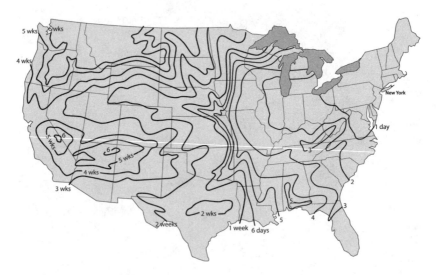

Figure 19

In 1857, the time needed to travel decreased considerably from that of a few decades earlier. (After *Atlas of Historical Geography of the United States*, used with permission)

Figure 20
With the arrival of Christmas, postal workers became overloaded with packages as this holiday transformed into a gift-giving occasion.

Figure 21
Sir Edward M. Pakenham, commander of the British troops, fought against Andrew Jackson on a Louisiana battlefield.

Figure 22
Andrew Jackson, commander of the US troops, was positioned at Chalmette plantation, located a few miles south of New Orleans.

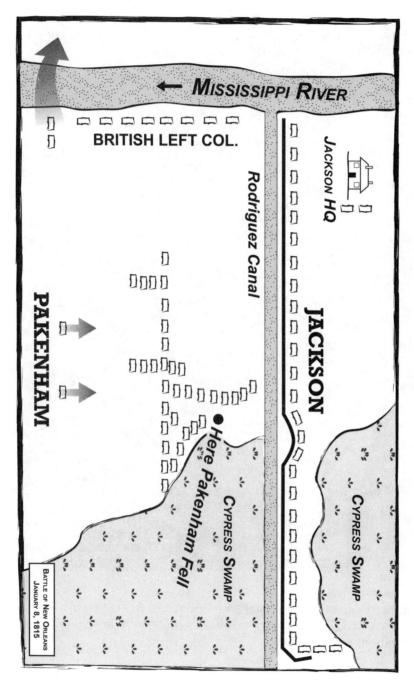

Figure 23

A map depicting the Battle of New Orleans on January 8, 1815.

Figure 24

On the deck of the *Sully*. As Samuel F. B. Morse traveled back to New York from his years in Europe, he came up with the idea of using electricity to send messages by compressing words into a code.

Figure 25
Morse rushed to the gravesite of his wife, Lucretia, who was buried in a family plot in New Haven, CT. Her death later inspired the creation of the telegraph.

Figure 26
Samuel F. B. Morse invented a way to communicate rapidly with his electromagnetic telegraph.

Figure 27
Alfred Vail brought to life, and often improved, many of Morse's ideas.

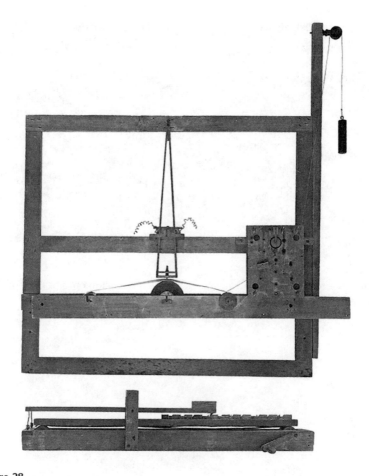

Figure 28

Morse's early telegraph was made from parts in his studio. The code was received by an electromagnet mounted on a canvas frame, which pushed a pencil in and out to write the code onto a strip of paper. A see-saw device rode over a set of teeth transmitting the code.

Figure 29
Moments after President James A. Garfield entered the Baltimore and Potomac Railroad Station in DC he was shot.

Figure 30
Charles Guiteau was an unstable man who assassinated President Garfield.

Figure 31
James A. Garfield was the beloved twentieth president of the United States.

Figure 32
On hearing the news of the shooting, Lucretia Garfield rushed to be at her husband's side.

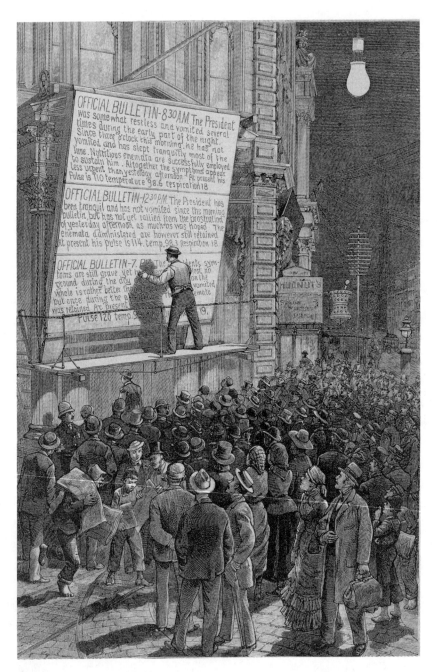

Figure 33
Crowds in New York City learned about Garfield's condition by reading messages on bulletin boards originating from telegraphed dispatches from the White House.

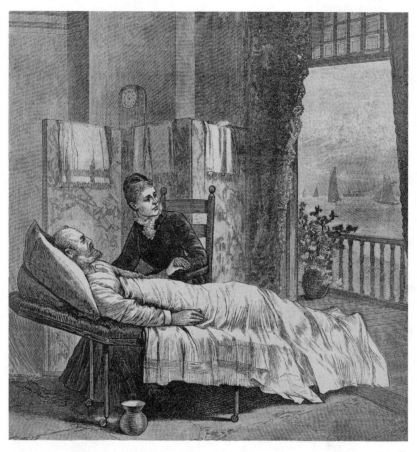

Figure 34
As the ailing Garfield moved closer to the oceanside he loved, his wife never left his bedside—neither did the country, by way of the telegraph.

Figure 35

Leland Stanford, an American magnate, funded Eadweard Muybridge's photography work in order to answer a question about how a horse gallops.

Figure 36

Eadweard Muybridge was a photographer who ushered in an age of capturing motion with a series of cameras.

Figure 37

When a horse gallops, there is an instant when all of its hooves are off the ground. Muybridge provided a picture to answer Stanford's question.

Figure 38

A segment of the Palo Alto racetrack was flanked on one side by a shed of cameras and by a tilted backdrop to provide more light on the other side. As the horse ran, its body stretched a thread going across the track, which triggered the camera's shutter and snapped a picture.

Figure 39

The camera's shutter, released quickly using electricity, created a peek-a-boo effect with the camera lenses. This was part of Muybridge's secret for successfully snapping a picture of a horse in midair.

Figure 40
Hannibal Goodwin was a Newark, NJ, preacher who desired to make pictures for his Sunday school lessons. He invented a flexible camera film using chemistry.

Figure 41
George Eastman, a photography entrepreneur, was involved in a long legal battle with Hannibal Goodwin over determining the original inventor of flexible camera film.

Figure 42
Hannibal Goodwin lived at the Plume House located next to the House of Prayer Church, in Newark, NJ. He had a chemistry laboratory in the attic, where he made his camera film.

Figure 43
Reverend Goodwin sawed a five-foot hole into his attic's roof to bring daylight into his chemistry laboratory.

Figure 44
Frederick Douglass, a lecturer and abolitionist, was once the most photographed man in the world and used his likeness to offset the stereotypes about blacks.

Figure 45
W. E. B. DuBois, a scholar of African-American studies, believed that commercial camera film did a poor job of depicting black skin.

Figure 46
Caroline Hunter began the Polaroid Revolutionary Workers Movement (PRWM) with Ken Williams to bring to light the nefarious use of her employer's instant photographs within South Africa's apartheid system.

Figure 47
A Polaroid ID-2 camera similar to this model was used to make photographs of black South Africans for their passbooks, allowing the country to control their whereabouts.

Figure 48
William Wallace demonstrated his electric arc lamp to Edison in Ansonia, CT, providing a catalyst for Edison's efforts to make an electric light.

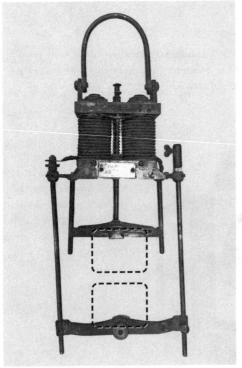

Figure 49
An arc light made by William Wallace. (The dashed lines indicate the position of the carbon blocks, which provided illumination.)

Figure 50

Wallace's telemachon converted water power from the Naugatuck River into electricity.

Figure 51

Wallace tested his arc light by attaching one to the chimney of the Wallace factory, illuminating the city and stirring up the townspeople.

Figure 52
Edison's laboratory in Menlo Park was a beehive of activity day and night.

Figure 53
Edison (center) and his men momentarily pause from their work on the second floor of Edison's Laboratory.

Figure 54
A young Thomas Edison.

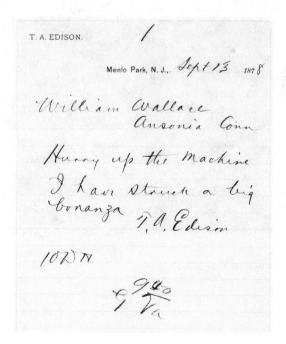

Figure 55
A letter from Edison to William Wallace urging the Connecticut inventor to hurry with his delivery.

Figure 56
One of Edison's earliest light bulbs.

Figure 57
NASA's John Casani with a Golden Record before it was bolted onto the Voyager spacecraft.

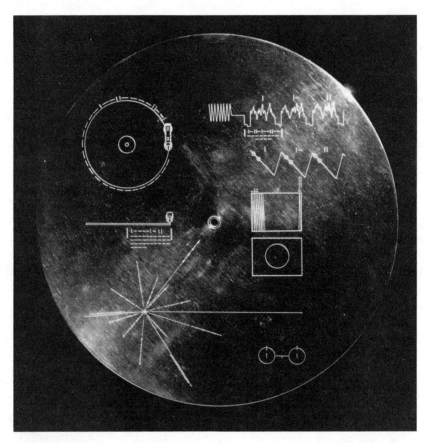

Figure 58
The Golden Record had directions inscribed on the album cover showing aliens how to use it.

Figure 59

Edison's phonograph was able to capture sound by pricking tin foil that was wrapped around a cylinder.

Figure 60
A small boy with a phonograph in his cabin home, showing how the ability to listen to music was democratized.

Figure 61
The cassette tape enabled listeners to share and record personalized mix tapes.

Figure 62

Jacob Hagopian was an engineer who helped to change the shape of data by creating the magnetic layer for the early IBM hard disk.

Figure 63

Rey Johnson took on the mission to find a way to store data without IBM's punchcards.

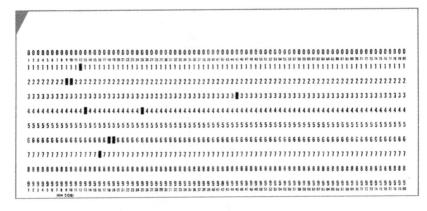

Figure 64

A punchcard held information by the position of its holes, but the number of cards was becoming unmanageable.

Figure 65

Herman Hollerith found a way to collect and tabulate census data by punching holes into cards.

Figure 66

This woven picture of Joseph-Marie Jacquard was made by a loom taking instructions from cards with holes. (Holes in cards allowed needles to pass and build the picture.)

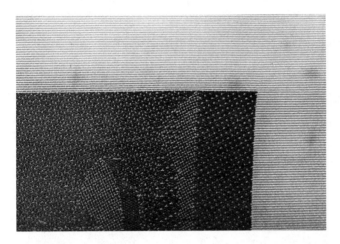

Figure 67
A magnified view of the surface of the Jacquard portrait showing that it is a woven fabric.

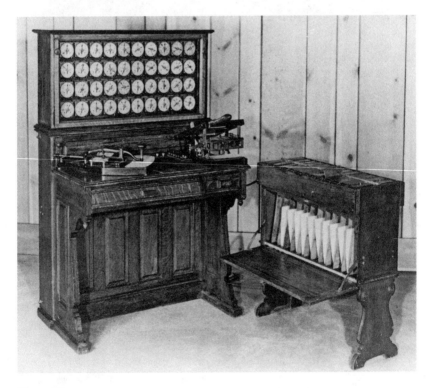

Figure 68
Hollerith's machines punched, tabulated, and sorted cards full of data.

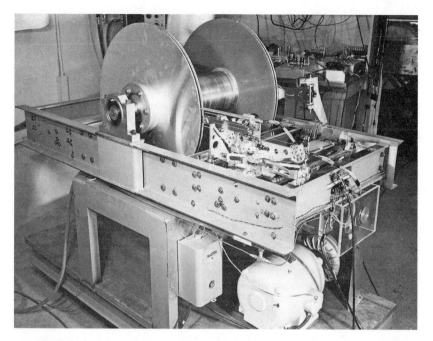

Figure 69
The earliest IBM hard disk was made up of scrap parts.

Figure 70
These three pictures show how Hagopian spread the magnetic coating by spinning the disk.

Figure 71
The RAMAC, IBM's commercial hard disk, stored 5 MB of data.

Figure 72
The IBM RAMAC took several men to prepare it for shipment.

Figure 73
The birthplace of the hard disk is 99 Notre Dame Avenue in San Jose, CA.

Figure 74

A street view of the London hospital where penicillin was discovered. Fleming's lab faced the street and was located at the second window above the round street-level plaque in the nearest part of the building.

Figure 75
Inside Alexander Fleming's laboratory at St. Mary's Hospital in London.

Figure 76
Alexander Fleming sitting at his microscope around the time of his discovery
of penicillin.

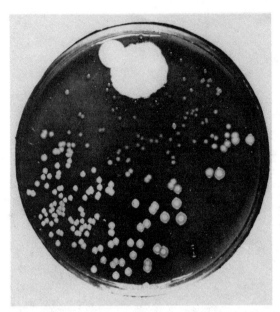

Figure 77
The original Petri dish that Fleming found containing the mold that produced penicillin.

Figure 78
German chemist Otto Schott, who invented a new glass called borosilicate glass, which is widely used in science laboratories.

Figure 79
Ernst Abbe, a German scientist, worked with Schott to improve the poor quality of glass lenses and glassware used in scientific endeavors.

Figure 80
Microscopes labeled "JENA" were highly desired for their German-made high-quality glass lenses.

Figure 81
Jesse and Bessie Littleton helped birth Pyrex. Bessie desired a shatter-proof cooking dish. Her husband, a Corning glass physicist, brought home glass dishes for her to test.

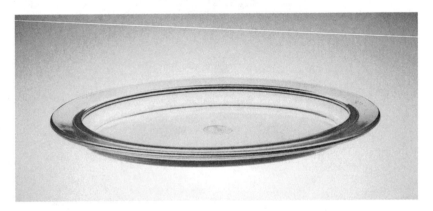

Figure 82
A Pyrex dish got its strength from its ingredients, particularly the element boron.

Figure 83
A Pyrex beaker was able to hold hot liquids and even acids because it was made of a new form of glass.

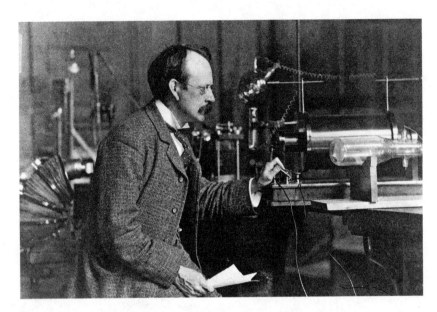

Figure 84
Sir J. J. Thomson peering into his glass tube at his Cambridge University laboratory.

Figure 85
Ebeneezer Everett, a deft technician, brought J. J. Thomson's ideas to life.

Figure 86

The scientific glass tube, crafted by Ebeneezer Everett, helped J. J. Thomson observe the behavior of the cathode ray—and discover the electron.

Figure 87

Phineas Gage was a railroad foreman whose unfortunate accident with a tamping rod has informed neuroscientists about the workings of the brain. (Note: This daguerreotype picture is a mirror image of Gage.)

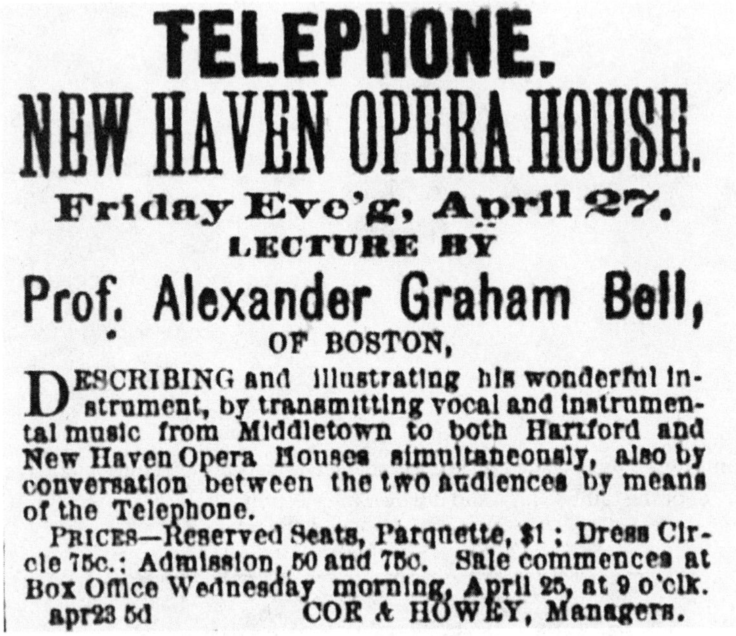

TELEPHONE.
NEW HAVEN OPERA HOUSE.
Friday Eve'g, April 27.
LECTURE BY
Prof. Alexander Graham Bell,
OF BOSTON,

DESCRIBING and illustrating his wonderful instrument, by transmitting vocal and instrumental music from Middletown to both Hartford and New Haven Opera Houses simultaneously, also by conversation between the two audiences by means of the Telephone.

PRICES—Reserved Seats, Parquette, $1 ; Dress Circle 75c.: Admission, 50 and 75c. Sale commences at Box Office Wednesday morning, April 25, at 9 o'clk.

apr23 5d   COE & HOWEY, Managers.

Figure 88

The *New Haven Evening Register* announced a "telephone" concert in 1877.

Figure 89

Alexander Graham Bell, the inventor of the telephone, demonstrated his creation to a New Haven audience.

Figure 90

Bell spoke into an early telephone like this on a New Haven stage.

Figure 91

George Coy started a telephone exchange company in New Haven, CT, using the telephone license he obtained from Bell.

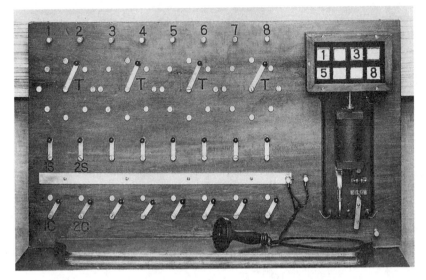

Figure 92
Coy's switchboard used carriage bolts as the end of telephone lines, which were connected with levers moved with teapot handles. On the back side, thin wires originating from Mrs. Coy's undergarments made up the electrical circuitry.

Figure 93
The first telephone exchange was located on the ground floor of the Boardman Building in New Haven, on the street facing forward and located near the corner.

LIST OF SUBSCRIBERS.

New Haven District Telephone Company,

OFFICE 219 CHAPEL STREET.

February 21, 1878.

Residences.

Rev. JOHN E. TODD.
J. B. CARRINGTON.
H. B. BIGELOW.
C. W. SCRANTON.
GEORGE W. COY.
G. L. FERRIS.
H. P. FROST.
M. F. TYLER.
I. H. BROMLEY.
GEO. E. THOMPSON.
WALTER LEWIS.

Physicians.

DR. E. L. R. THOMPSON.
DR. A. E. WINCHELL.
DR. C. S. THOMSON, Fair Haven.

Dentists.

DR. E. S. GAYLORD.
DR. R. F. BURWELL.

Miscellaneous.

REGISTER PUBLISHING CO.
POLICE OFFICE.
POST OFFICE.
MERCANTILE CLUB.
QUINNIPIAC CLUB.
F. V. McDONALD, Yale News.
SMEDLEY BROS. & CO.
M. F. TYLER, Law Chambers.

Stores, Factories, &c.

O. A. DORMAN.
STONE & CHIDSEY.
NEW HAVEN FLOUR CO. State St.
"   "   "   " Cong. ave.
"   "   "   " Grand St.
"   "   "   Fair Haven.
ENGLISH & MERSICK.
NEW HAVEN FOLDING CHAIR CO.
H. HOOKER & CO.
W. A. ENSIGN & SON.
H. B. BIGELOW & CO.
C. COWLES & CO.
C. S. MERSICK & CO.
SPENCER & MATTHEWS.
PAUL ROESSLER.
E. S. WHEELER & CO.
ROLLING MILL CO.
APOTHECARIES HALL.
E. A. GESSNER.
AMERICAN TEA CO.

Meat & Fish Markets.

W. H. HITCHINGS, City Market.
GEO. E. LUM,   "   "
A. FOOTE & CO.
STRONG, HART & CO.

Hack and Boarding Stables.

CRUTTENDEN & CARTER.
BARKER & RANSOM.

Office open from 6 A. M. to 2 A. M.

After March 1st, this Office will be open all night.

Figure 94

Coy initially had twenty-one customers in his first week of operation and created this early telephone directory.

Figure 95
Almon Strowger was an undertaker whose strong dislike for female telephone operators led to his creation of an automated telephone exchange.

Figure 96
With early telephones, calls were connected by female switchboard operators who were known as "hello girls."

Figure 97

Strowger's invention of an automatic telephone exchange was based on pins fanned out in a circle that could be touched by a mechanical finger in the center to connect a call.

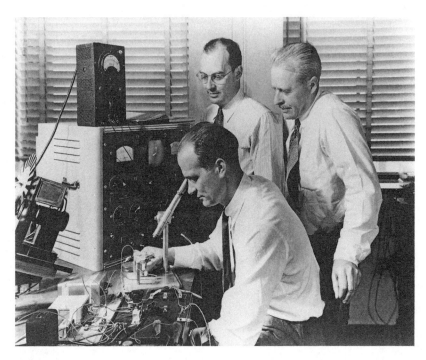

Figure 98
Walter Brattain (right) and John Bardeen (left), inventors of the transistor, standing behind their boss, William Shockley, who is sitting at their microscope.

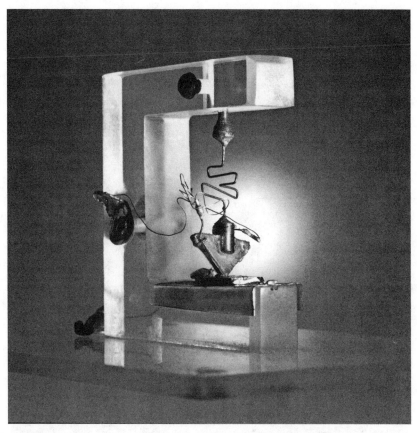

Figure 99

In 1947, Bell Labs scientists created the transistor, which allowed the signals of telephone calls to be switched as well as amplified over long distances across the nation.

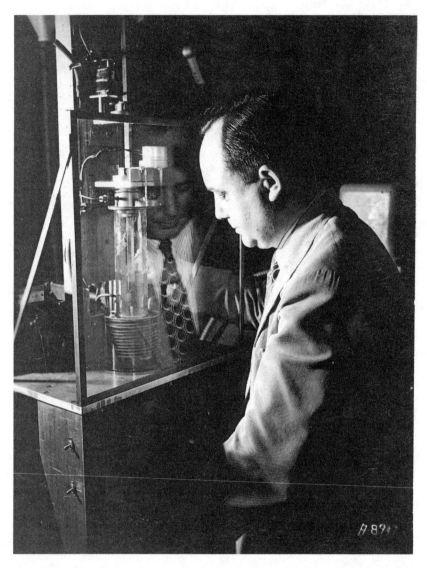

Figure 100
Gordon Teal, a Bell Labs chemist, pulling a pristine germanium crystal from a small molten pool.

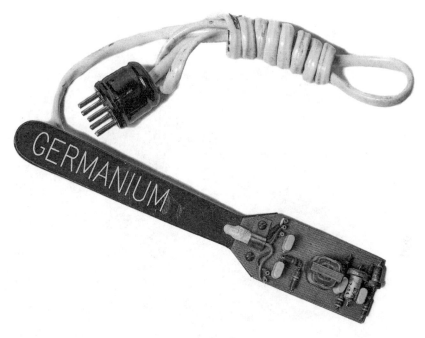

Figure 101

When Gordon Teal dipped this electronic paddle into hot oil, the phonograph connected to it stopped playing, demonstrating the flaw of germanium transistors.

Figure 102
Small transistors like these were in Gordon Teal's pockets during his presentation when he announced that the age of silicon electronics had begun.

6
Share

How magnetic bits of data made it possible to share, but also made it difficult to stop what is being shared about us.

NASA's Out-of-This-World Record Album

In 1977, as Steven Spielberg was putting the finishing touches on *Close Encounters of the Third Kind*, a movie in which humans use musical notes to communicate with aliens, NASA was preparing to send its own message to share with extraterrestrials, too. That year, the space agency had a once-in-a-lifetime opportunity with the launching of its two Voyager spacecrafts, for they would be able to travel faster and farther than originally conceived. The planets were lining up in a unique geometry that happened only every 176 years, and in this formation, one planet could toss a spacecraft to the next planet and then the next, like a hot potato. By using the gravity of the planets, these space vehicles could be slingshot across the solar system, using less fuel but acquiring great speeds, reaching the farthest of distances and possibly alien worlds.

Accompanying these Voyager spacecrafts would be a message, but not just an ordinary one. The contents would be historic, representing a culture like an early map or a carving on a cave wall. The magnitude of the message was great because calculations showed that the Voyager crafts would be able to travel unimpeded for billions of years, outliving Earth itself, which other calculations predicted would be swallowed by the sun in less than that time. With that, these twin Voyager spaceships were promoted from being just space probes to being the last artifacts of human life, carrying a parcel of Earth's last data.

The idea for the space message culminated a year earlier, in 1976, when John Casani, the project manager of the Voyager mission, contacted Cornell astronomy professor Carl Sagan around Thanksgiving and asked him to come up with some kind of message to attach to the spacecrafts. Sagan said, "Absolutely."

Sagan decided on sending a record. Unlike a vinyl disc, a recording medium that was popular on Earth in the 1970s, this record was going to be a twelve-inch gold-coated copper disc. The Golden Record, as it was called, for each of the twin Voyager rocket launches would contain greetings from Earth, images, sounds, and music. Sagan pulled together friends to form an ad hoc Voyager Record Committee, which included his wife, Linda Salzman Sagan; Jon Lomberg, the illustrator for Sagan's books; Timothy Ferris, a writer at *Rolling Stone* magazine; and, Ann Druyan, a novelist and Ferris's fiancé. Each member handled different parts of the record's contents, but all of them contributed to the music.

Selecting the music for the ninety minutes of playing time designated to represent the entire Earth had technical challenges as well as human ones. Before the age of digital files, music resided on physical discs and cassettes, which had to be located in the bins at Tower Records and other music stores and then be manually carried to the studio to be played. In addition to these technical difficulties, was the challenge of selecting what to send into space. Music selections touched the personal, such that individual taste became a guiding force, unlike the impersonal math that made the space trajectories possible. For members of the Voyager Record Committee, being Noah for a space ark strummed a human chord. Unknowingly, their judgment for musical selections was being clouded by bias, too.

Deciding what music to send from Earth was a thought experiment that had already been considered in a best-selling book. Years earlier, in 1974, a respected scientist named Lewis Thomas proposed, "I would vote for Bach, all of Bach, streamed out into space, over and over again," which he wrote in *The Lives of Cells*. "We would be bragging, of course, but it is surely excusable for us to put the best possible face on at the beginning of such an acquaintance," Thomas wrote. "We can tell the harder truths later." The original selections for the Golden Record mirrored this best-selling book's mindset, barely representing the whole planet. Most of the songs were classical music, a genre of which Sagan was a fan, originating from a small region of Europe, and not the whole blue pale dot, as Sagan loved to call Earth. Slowly, however, the selections began to include music from other cultures. With the urging of the younger members of the team, the suggestions from an anthropologist, and the chiding and prodding by legendary song hunter Alan Lomax, minds were swayed and the playlist began to reflect the entire planet. Soon, the Golden Record was a true sampling of the place from where it came. Beethoven's Fifth Symphony, with its iconic first notes that rip open the silence of space, was also accompanied by Senegalese percussion, Azerbaijani flutes, Navajo chants, Melanesian panpipes, and African-American jazz.

With one Golden Record blasting from Earth on August 20 and the other on September 5th of 1977, their long journeys as Earth's mixtapes began. NASA's mission had the initial intention of collecting data about space, but it also dispatched data—the world's music.

This event in 1977 was made possible by an invention created exactly one hundred years earlier—the phonograph. In 1877, Thomas Edison, on a fluke, created a contraption that would be important to society, for it made possible not only the storing of music, but also the ability to share it. Edison tapped into an old love and tradition, for music is important to most cultures.

Today, our modern sensibilities cannot imagine a world where music is not available on the ready, but it once wasn't. In order for music to be easily accessible, it had to go through a metamorphosis in Edison's day. Music had to change its shape. It had to become physical. Music had to become data.

Before 1877, no machine could record and play back a human voice. As such, the pitch and cadence of anyone who died before Edison's innovation were unknowable. Generations would never be familiar with Confucius's voice or Shakespeare's. Generations would never know the sound of Abraham Lincoln's voice or Frederick Douglass's. Generations would never learn how Poe or Dickinson actually read their work. Even the pronunciation of ancient languages, like the spoken language represented by Egyptian hieroglyphics, would elude scholars forever. The capturing of sound before the nineteenth century was a formidable task, a dream parallel to lassoing light or bottling a breeze. The poet Ralph Waldo Emerson foresaw Edison's technology, when he wrote, "we shall organize the echo." But in 1877, Edison did more than just organize echoes. He made them tangible and portable and able to be played back.

Edison's Sonic Dream

In the summer of 1877, thirty-one-year-old Thomas Edison was catapulting technologies of the nineteenth century into the future and had his eyes set on two inventions. In his lab and in his musings, he wanted to make a way to automatically write telegraph messages from Samuel Morse's invention, and he wanted to fix a flaw in Alexander Graham Bell's telephone. Edison was a master at refining existing inventions and it wasn't uncommon for him to juggle several ideas at once. It was while working on both the telephone and telegraph on the ordinary day of July 17, 1877, that the idea of combining them, like peanut butter and chocolate, came to him. By merging the ability to write from the telegraph and the capacity to receive sound from the telephone, Edison devised what he later called his favorite creation, a machine with the capability to write sound. He called it the phonograph.

The summer months of 1877 were hectic as Edison tried to make a better telephone to catch up with Bell's popular product invented the year before, as well as keep pace with the flood of ideas from his own head. One section of his long-roomed laboratory contained tables full of springs, levers, and sharp tips to make a machine that pricked dots and dashes into strips of paper with special coatings as a way to record messages from Morse's telegraph. In another part of the room, Edison was also experimenting with the telephone. Although Alexander Graham Bell had beaten him, Bell's design had a problem. Whenever words with consonants of t, p, v, and c were spoken, they hissed; the sounds "s," "th," and "sh" were inaudible. Every day, Edison's workers found him yelling into a cone-shaped mouthpiece with his fingers on the back, feeling the vibrations in the thin material over the narrow part of the cone, called a diaphragm. Edison tested several candidates for this material to find which one faithfully trembled with a human voice. His notebooks swelled with drawings, as his entries pitched between the telephone and the telegraph. It was in the course of these hot and humid days of intense thinking, surrounded by parts that pricked paper and slices of resonating slivers, that his idea hatched.

During an ordinary midnight dinner, when the activity in the beehive of Edison's laboratory momentarily paused, the Wizard of Menlo Park, his shaggy hair pointing in all directions, was still working on his quivering materials. Thinking out loud, he stated an idea to his chief assistant, Charles Batchelor, with his legendary overconfidence. "If we put a point on the centre of that diaphragm and talked to it whilst we pulled the wax paper underneath," he said, "it would give us back talking when we pulled the paper through the second time." His idea hit everyone in the laboratory like a thunderclap. Capturing the human voice and then listening to it was a thrilling proposition because no such thing existed. On hearing Edison's words, the rapid scouring for parts to make a talking machine began, as if within earshot of a start gun.

Instruments resting on the wooden tables from some of Edison's earlier projects were immediately pilfered and repurposed. One person cut off the sharp tip of a needle and soldered it to a circular diaphragm. Another person fastened the diaphragm and mouthpiece to a wooden stand. Yet another cut a strip of wax-coated paper and placed it under the diaphragm's needle. In less than an hour, an apparatus appeared before the wizard. The room quieted as Edison sat down, leaned his portly frame forward, and nested his lips onto the mouthpiece. He then yelled "Halloo," as his assistant Batchelor pulled the strip of waxed paper underneath slowly and evenly, like fishing line in a pond. When Edison stopped shouting into the horn, he and Batchelor looked at the paper and noticed the line was wide and then narrow, like a digesting earthworm's figure. The two put the paper back where it started, and pulled it under the diaphragm again. "I listened breathlessly," said Edison. "We heard a distinct sound," he said, "which a strong imagination might have translated into the original 'Halloo!'" The nearly deaf Edison heard something, but Batchelor was skeptical.

The seeds had been planted for the development of a talking machine, or phonograph, but it would have to wait. Edison was being pulled back to his telephone and telegraph projects, and was starting to look into a new form of illumination in electric lights. Months passed, and even though Edison could not get back to the phonograph, he kept drawing designs in his notebook. At the end of November, he found time to think about his machine, deciding on using a cylinder to store a voice after considering both a disc and a long strip of paper. The genius of his design was its simplicity: A mouthpiece collected the sound waves, which pushed the diaphragm, like a trampoline, and a sharp tip attached to the diaphragm moved up and down, pricking the tinfoil wrapped around a cylinder. After much thought and a few different iterations, the Thursday after Thanksgiving Edison sketched his design and passed it to his trusty machinist, John Kruesi, and told him of his intention to make a machine that talked. Kruesi looked at him incredulously.

Kruesi, in a machining marathon, spent the first six days of December making the phonograph. While bringing Edison's idea to life, Kruesi engraved a spiral groove around the brass cylinder, like a candy-cane stripe, to provide a track to guide the needle as it traveled, and also to make legroom for the sharp tip to push into the foil. He, along with Charles Batchelor, attached the tinfoil to the cylinder, before passing it on to their boss for testing on December 6. The wizard then hovered his lips over the mouthpiece and prepared to give his brainchild its first words. Uttering what Edison often said to his young children, nicknamed "Dot" and "Dash," he shouted "Mary Had a Little Lamb." While this nursery rhyme was not prophetic like Morse's "What Hath God Wrought" in 1844, it was certainly more intentional than Alexander Graham Bell's "Watson! Come here. I need you," a year earlier in 1876. When another speaker cone was attached and the crank was turned, Edison's words came out faintly, but undeniably. "I was never so taken aback in my life," he said.

Admittedly, his invention was flawed. The words "Mary had a little lamb" probably sounded like "ary ad ell am" on the first try. Additionally, the phonograph could hold less than a minute of sound, limited by the length of the helical groove on the cylinder; and, because of the softness of tin, the message could only be played two or three times before the metal deformed and eroded the sound beyond the point of being decipherable. Nevertheless, Edison's enthusiasm never waned as he and his men toiled all night in making the phonograph as articulate as possible. They wanted to show their creation to the world the next day.

On December 7, 1877, Edison and Batchelor boarded the train from the tiny wood platform in Menlo Park, NJ, to head to New York City and were joined by Edison's business associate Edward Johnson to visit the offices of *Scientific American,* the premiere source of science news. There, they placed the phonograph on the editor's desk as a few onlookers gathered. Edison turned the crank, and the growing crowd, making the floor creak, heard "Good Morning. How are you doing? How do you like the phonograph?" and then the phonograph bid the crowd a good night. *Scientific American* did something it rarely did. It stopped the presses that day to alert all of humanity that life had changed. "Speech has become," they said, "immortal."

Edison created a new way to represent information in addition to written language. Words on a page had two lives—a spoken existence and a written one. But sound only had one life to live. Sound was confined to a short time span and a habitat confined between one person's lips and another's ears. Beyond these bounds, sound, like a snowflake, left no signature. For these reasons, when Edison spoke "Mary Had a Little Lamb" into his phonograph, his words were a milestone of human progress equivalent to Neil Armstrong's "one small step for man, one giant leap for mankind," while stepping onto the moon. With the phonograph, utterances, like a baby's first words, could be heard and cherished at any time. Without Edison's—or humanity's—realizing it, however, Edison changed the shape of data. Information had undergone a metamorphosis from scribbles on parchment and words stamped on paper with Gutenberg's printing press to Edison's pricks in tin.

The Wizard of Menlo Park had a vision for his favorite invention and a few months after he created it, he made a list of predictions for the uses of his brainchild. They included audiobooks, educational lessons, last testimonies, music, toys, and answering machines, many of which exist today. Edison also believed the primary use for his invention would be dictation for business. Here, he was wrong, for where the phonograph would leave its mark was in music.

Before the phonograph, the diffusion of songs was made possible by live performances of traveling singing troupes or with sheet music played by local talent. The phonograph ignited the nation's imagination and would soon be found in the furthest outskirts of civilization, from the plushest parlor of wealthy homes to the most decrepit houses of the poorest farmers—democratizing the listening of music. Edison's dream was that with his phonograph, any person from any walk of life could have a song. And, with his invention, everyone did.

The wizard brought music into people's lives and before long how society experienced music changed with his phonograph. During a live performance, music was a shared event in a concert hall, or park, or juke joint between performers and the audience, and between audience members with each other. With the phonograph, the collective musical experience shrunk from large halls to living rooms, but the trade-off was that the music could now be played at any time. The phonograph was one of Edison's favorite inventions, but not everyone was a fan of Edison's phonograph. John Philip Sousa, the patron saint for marching bands, believed the phonograph would bring "a marked deterioration in American music and musical taste." Yet the sales of phonographs grew. By 1906, thirty years after Edison's invention, over twenty-six million records were sold. Fifty years later, by 1927, record sales were a hundred million.

The public found music from the phonograph to be irresistible, but they might not have known that the phonograph was shaping the music they enjoyed along the way. Just as Alexander Graham Bell's early telephone could not pick up sounds like "s" and "sh," Edison's phonograph was similarly limited. Cellos, violins, and guitars produced tones too soft for the early phonograph to pick up, so louder instruments like pianos, banjoes, xylophones, tubas, trumpets, and trombones became preferred for recordings of music. Additionally, phonographs helped to fashion musical styles in a nation that was highly segregated. Blacks and whites did not socialize, but phonograph records crossed these racial divides, enabling white and black musicians to hear and borrow styles from one another. Phonographs were dispatches of culture. This sharing of music between these musicians helped fashion jazz and blues and later rock and roll, creating a cohesion of society that Edison could never have predicted.

One hundred years after the birth of the phonograph, in 1977, the progeny of Edison's invention was continuously evolving. One branch of the family tree produced the record using analog grooves to hold data; another branch was the cassette tape, using magnetic bits to register notes. Each had its shortcomings. Records were bulky, but a listener could get to a song without delay, and the musical reproduction quality was high. Cassette tapes were pocket-sized, but getting to a song took patience, and the sound quality was relatively limited. Like all cousins, they would look very different from each other and from their common ancestor, but the family trait of being a vehicle to share and disseminate music remained.

Edison's creation of the phonograph in 1877 eventually made it possible for music to be purchased in stores. By 1977, cassette tapes would only amplify the frenzy to buy and borrow, consume and collect music. But this offspring would have a new family trait. A blanket of magnetic dust stuck on a strip of plastic within cassette tapes not only made it possible for a person to hear tunes, but also gave them the capacity to copy sounds for themselves. This recording feature enabled listeners the freedom to curate music based on their own personal preferences, and this ability to collect, copy, and curate music gave rise to the mixtape (the ancestor of the playlist).

The mixtape allowed listeners to personalize sounds. The contents of the mixtape, starting in the late 1970s, when polyester pants were in vogue, defined a person's mood, their thoughts, their concerns, and their situation. Since the 1970s, the mixtape became a token of affection, a gift of friendship, and a sign of love. The music represented the best of the giver, or what they aspired to be. With this new superpower of selecting and organizing music that was meaningful to a listener, the mixtape, in a way, became their sonic incarnation. The mixtape became them.

Mixtapes and prerecorded cassette tapes helped the spreading and sharing of music in many ways. A cassette tape in a boombox shared music with anyone within earshot. A cassette tape gave musicians a means to share their music outside of the established distribution channels of the music industry with the demotape. A cassette tape in a Sony Walkman, the 1980s version of the iPod, gave listeners the ability to share music with themselves in their own protective musical bubble. The orchestrating of magnetic bits of iron within the over 130 million cassette tapes sold in 1977, the year of the Golden Record, further democratized music as Edison's phonograph had done one hundred years earlier.

As society shared music and mixtapes with total abandon, what went unnoticed in the leap from Edison's phonograph to the magnetic cassette tape was that the shape of data changed, too. The phonograph's cylinder, and later disc records, had surfaces covered with grooves that looked like hills and valleys, matching the thrust of the sound waves that created them. In digital recordings, the sound wave was translated into electricity to convince magnetic patches on the tape to be weak or strong magnets, speaking the language of binary of ones and zeros. Society moved from analog grooves in foil and wax to digital magnetic bits. While people were busily copying sounds from the radio and from their favorite albums, the world was entering the age of binary, as the shape of data changed.

This step was significant because binary is the language of computers. As more devices spoke in binary, they brought the reality of more machines talking to each other and the greater possibility of a world of automation, eventually allowing computers to think.

Binary seems like a modern concept; however, twenty years before Edison was working on his analog phonograph in 1877, the seeds for making the modern world digital had been planted by mathematician George Boole in Ireland. In 1854, Boole, who loved languages, found that simple statements of logic could be represented with a symbol, and their relation to each other could be established with a true and false value. Eighty years later, an MIT graduate student named Claude Shannon applied Boole's esoteric math theorem to circuit switches that went on and off, giving his machine an ability to compute and think. Claude established a language for computers and in order for machines to work together, all information had to be reduced to these basic units or "bits" of "1" or "0," including information that was music. Once a device became digital, the need for humans was less important. The machine could do the task on its own.

While few books or newspapers mention it, the conversion of data to a magnetic form was a milestone, for it completed the age-old desire for fitting more information into a smaller space. In addition, the switch to magnetic data removed the need for humans, since data was processable by computers with binary. Unknowingly, the digital form also made it possible for data, such as music, to be able to peel away from its physical container in order to be streamed from our devices. Music we enjoy from streaming sites and websites doesn't come from the beautiful facade on our screens, but from unattractive buildings or data centers full of hard disks. Our data is not just a result of a click, but the action of magnetic bits. Before those large warehouses of information—and our music—were possible, though, the hard disk had to be born. That birth required coaxing magnetic dust to behave.

The West Coast Boys

In the summer of 1952, Jacob (Jake) Hagopian, a thin, energetic, precisely groomed engineer of Armenian descent, was the thirty-third employee hired to join IBM's west coast laboratory in San Jose, California. He applied for a new job he saw in a local newspaper touting an "exceptional opportunity," but the work he was going to do wasn't clear. IBM wanted Californian engineers, but the cold weather of its headquarters on the east coast was no draw, so Big Blue, as IBM was called, was setting up shop to take advantage of the creative talent out west. Hagopian was now an IBM-er, and was brought on as a consulting engineer, which, like an internal min-uteman, allowed him to work on the most pressing problems. This suited him perfectly, for Hagopian was a highly experienced engi-neer, with a knack for breaking down problems into understandable chunks. This ability was exactly what his new boss needed.

Hagopian's boss was Reynold (Rey) Johnson, a tall, redheaded, farm-raised Minnesotan of Swedish stock, whose hands swallowed others in a handshake. Johnson was thrust into this west coast ven-ture just months prior. One winter afternoon in January 1952, IBM management asked Johnson to move from IBM's Endicott campus in New York to California and uproot his family. Johnson was a decade shy of joining IBM's Quarter Century Club, for twenty-five years of service, and easing into a slower pace from the comfort of his upstate New York home. His boss had other plans.

IBM had a problem. IBM was producing sixteen billion punch-cards a year. This growth was not sustainable, because it was becoming too difficult to store, sort, and manage all those cards. Punchcards were spawned from the need to count millions of citizens for the census, which was originally tabulated by hand. Herman Hollerith created the punchcard by putting holes in certain regions of the card to provide information. He got the idea from two places. At the end of the nineteenth century, train conductors punched tickets that described the rider's physical features. Hollerith borrowed that idea. Also, in the 1800s, Joseph Marie Jacquard invented a loom that made intricate weave patterns from instructions on thick cards with holes in them. These holes allowed long weaving hooks, attached to threads, to rain down on flat fibers, stitching the image layer by layer. Where holes were present, threads could pass. Where holes were absent, stitches were blocked. The use of holes to provide information was the center of Hollerith's invention and, with it, data transformed from words to holes.

Before Hollerith's invention, the census in 1880 took nearly seven and a half years to complete. With the Hollerith system of machines that counted the holes, the US population of nearly sixty-five million took two months to tally twice in 1890. The ease brought by the new form of data was undeniable. Once counting was done, the data could be shared to help the government know itself—who are its citizens, what are the resources, what are the needs, what are the challenges. As more countries got counted, other countries wanted to be counted. The census provided the nation a mirror. Hollerith's business would be bought and folded into a newly formed company, called International Business Machines (IBM), bringing his punchcards to every corner of the world. But the punchcard was becoming a victim of its own success as their numbers grew. Now IBM had too many punchcards.

This Everest of punchcards brought Rey Johnson out to California. There was a problem to solve, but also an opportunity to do things a new way. IBM needed data to be stored compactly, which was not the case with stacks of punchcards, and for access to that data to be done in real time, automatically and instantly, which was also not the case with punchcard readers.

Inside IBM's new West Coast Research laboratory at 99 Notre Dame Avenue, Johnson was still deciding which direction to take for making a means to store data, but he was gaining clarity on the needs involved for that storage. Complaints from IBM customers said they wanted a way to arbitrarily access a transaction without going through all the punchcards. On January 16, 1953, Johnson called a small task-force meeting of the engineers to fix that punchcard problem. But the meeting was more profound than that. These men, with their white shirts, pocket protectors, and eyeglasses, were going to follow Edison's path and change the shape of data.

In the meeting, many voiced strong opinions of how information could be stored. One person proposed using a large magnetic cylinder, borrowing from Thomas Edison's idea for the phonograph, where Edison's rendition of "Mary Had a Little Lamb" arose from a needle running over a cylinder wrapped in tinfoil. In the magnetic version, a coating of magnetic iron replaced the tinfoil and a small magnet hovering above stood in for the needle. Another person proposed that a magnetic tape be used. Others suggested using magnets in the shape of sheets, rods, and even wires. For hours, occupants sat at a long Steelcase table deliberating on the shape of data, until one voice proposed a disk, like a record from a record player. That changed everything.

This idea was profound, for while a disk is geometrically simple, it provided an engineering advantage. Disks offered more area for music with an A-side and a B-side, allowing more data to be stored in a smaller amount of space. The same would hold for a hard disk.

Johnson's west coast team decided the first disks needed to be two feet wide, like large pizzas, with a gaggle of them spinning 1,200 times a minute, nearly twice as fast as a football spiral. They also agreed that the mechanism should resemble a jukebox, with disks stacked vertically, like books on a shelf. Now they needed to make it; to do so they headed to the junkyard.

Edison had always said that an inventor needed ideas and a big pile of junk, and these IBM engineers now had theirs. In the scrap heap, they found two metal girders, to support the spinning disks, that were heavy enough to prevent the hard disk from shuffling across the room, like a poorly loaded washing machine. To spin the disk, they found a motor. They also found an aluminum sheet. When the sheet was cut, it warped like a potato chip, so they flattened it with a tombstone from the graveyard.

The hard disk's mechanism was further inspired by the jukebox and the record player. In a record player, a stylus followed the pattern of the record's grooves, which served as the data that was turned into sound. On a hard disk, a layer of magnetic dust served as the medium to hold the data, for sound or any other information. The magnetic head flying above the hard disk substituted for the stylus, sensing the magnetic regions, which could be interpreted as ones and zeros—the basic units of the language of computers. It was Jake Hagopian's job to find a way to coat the disk with magnetic particles.

Coating the disk wasn't easy, since its thickness needed to be the same across a large area. Hagopian tried dipping the pizza-sized disk into a vat of paint, but the resulting surface was rough. He tried silk-screening the coating, but the surface was bumpy. He tried spray painting, but the surface was uneven. One day, while visiting a printing plant, he saw automated cylinders coated with ink spinning quickly to remove the excess. This put a seed in Hagopian's mind.

On November 10, 1953, Hagopian returned to the lab, picked up a twelve-inch disk, some paint, and a Dixie cup, and walked over to the machine shop. There he connected the disk to the drill to spin it and poured some of the paint in a ring in the center from the Dixie cup. When he started the drill, paint flew out in all directions, like spin art, hitting the newspaper Hagopian set around it. When the paint dried, he saw that it was the best coating he had ever produced—thin, uniform, and nearly flawless. To remove the lumps in the paint, Hagopian strained the paint through his wife's old silk stockings. Soon, spin coating became the official way to coat many of the early disks.

Next Hagopian had to figure out what magnetic particles to put into the hard disk's coating that would hold the data. First, he bought a bucket of a magnetic iron oxide dust from Minnesota Mining and Manufacturing (3M) at the high cost of $90 a gallon. He mixed the magnetic powder into a clear varnish and then spun-coated it onto a disk. The 3M product performed poorly and the coating easily scraped off with a fingernail. This was not going to work.

Hagopian looked to make coatings harder. One day, he saw an advertisement in *Life* magazine for a new form of unbreakable dinnerware called Melmac. Melmac was made of a hard plastic called melamine, manufactured by American Cyanamid, which came in a powder form. Hagopian bought this plate-ware plastic to make his soft magnetic coatings hard, tough, and smooth. It worked. But soon he was entering a realm that was beyond his training. He was going to need to get some help.

Hagopian needed magnetic particles and reached out to companies that use them for reasons different than his own. He called the California Ink Company, in San Francisco, which had a magnetic ink that was applied to the numbers at the bottom of bank checks so that they could be processed without a teller. Then he reached Ferro Enameling Company, a ceramic maker in Oakland, which used magnetic particles to add brown and black colors to their pottery glazes. Next he contacted the movie company Reeves Soundcraft Corporation, in New York City, which sold their iron oxide to movie studios to be applied to the edge of the film as the soundtrack. Finally, he wrote the paint company W. P. Fuller and

Company, in South San Francisco, which utilized iron oxide as a pigment to create orange and reddish colors, used for playgrounds and San Francisco Bay Area bridges. Here Hagopian struck gold.

The Fuller Paint company was more than willing to help Hagopian, making formulations for him in their lab by adding melamine to improve the hardness and also polyvinyl to make the red iron oxide paint pliable. They did so to a tune of $16 a gallon, which was a steal compared to 3M's $90. In the end, this paint worked.

It turned out that Fuller also made the iconic orange paint for the Golden Gate Bridge. Being a curious person, Hagopian ordered some of the bridge's paint and spun-coated the disk with its bright orange. The color was beautiful, but the magnetic field from it was too weak to store data. Testing the Golden Gate Bridge paint added some fun to the drudgery of Hagopian's work. He shared with his coworkers his experiments with the Golden Gate paint, but would live to regret it. Soon, a rumor spread that the hard disk's data layer came from the paint for the Golden Gate Bridge. "My complaint is that it trivializes [my work]," Hagopian said. The myth aside, Hagopian's efforts, along with others, would lead to the hard disk in computers, and soon the large data centers that hold the internet.

With the work of all the engineers over the years, all the pieces came together to make IBM's first commercial hard disk, the RAMAC—a random access method of accounting and control. The RAMAC was the size of two refrigerators, weighed over a ton, and held five million bits of data, or five megabits (about equal to one photo today).

The RAMAC was huge and didn't hold much data, but soon the data storage industry, which IBM helped to build, would follow the guiding principle of more data in less space. While silicon chips followed Moore's Law, the data industry doubled it. For every bit of real estate on a hard disk, there was a desire for smaller bits of data to live there. Less space would hold more information, and soon society developed a hunger for more data. This hunger would be fed with greater storage capacity for files, apps, games, pictures, and music, and consumers got accustomed to being able to share more. But the miniaturization of data would have other repercussions.

Music storage evolved from foil-wrapped cylinders to disc records to magnetic tape. But soon, music would shed its physical cocoon entirely, and like a butterfly, flutter across cyberspace as a digital file residing on a computer's hard disk, or in a MP3 player, or in data centers called the cloud. When music lost its shell and became a digital file, listeners benefitted from its being available at all times. The meaning for data changed, however. Data went from printed words to pricks in tinfoil to bumps on records to holes to magnetic bits to its disembodied form, and the evolution did not stop there. The ubiquity and small size of data on massive hard disks enabled the collection of colossal amounts of information about people. Music used to be the data that we collected, but now we are.

Music's new ethereal form changed how we experience it. Websites, starting with Napster and later YouTube, streaming services, social media platforms, and iTunes, brought music everywhere and to anyone, with downloads surpassing any of Edison's predictions. In addition, something occurred along the way that Edison might not have expected. The digital format shifted not only how we experience music, but what we share. Media services stream music to listeners, where music is the data, but data about the listener is gathered, too. These services know which song a listener chooses, how long the song is played, and how often, but they also amass data about where the listener is, when they listen, and who is around them. These websites and businesses, in turn, share this data they've collected about us with other businesses, agencies, and advertisers, all while we are simply enjoying music from our playlists.

With Edison's phonograph, music became the data to be collected, but with today's technology, humans are now the data. We have become the last step in the evolution of data, from Edison's recording of musical notes as pricks in tin to the tracking of our

7
Discover

How scientific glassware helped us discover new medicines and helped us discover the secret to our electronic age.

The Spoils of Science

Perched in his second-story laboratory at St. Mary's Hospital in London, Alexander Fleming looked into his microscope and pondered ways to fight disease. Sitting above the busy Praed Street in the fall of 1928, he was surrounded by glass in his small laboratory. Around him were pipettes, flasks, and petri dishes at a desk flanked by windowpanes. Here, tucked away in this red brick building, Fleming often returned to images of the Great War. During his tour of duty, ten years earlier, he saw scores of men survive the trenches, only to combat another foe: infection in an infirmary bed. Fleming witnessed that the battle against bacteria in the body was equally as lethal an enemy as human invaders in the field, for a severe burn or septic wound was often a death sentence. As soon as peace was declared and he was no longer needed for urgent care, he made it his life's work to help the body win its duel against microorganisms. This struggle against infection was an old war, with ancient scrolls prescribing bygone cures to battle what would later be called germs. Fleming tried to make a contribution to this longtime medical campaign, armed with his scientific glassware. His work was solid, but nothing spectacular, until a speck of dust changed everything.

With his mesmerizing blue eyes, graying hair, and large nose, this short, slight, and soft-spoken Scottish bacteriologist could have been mistaken for a wizard. While magic was not his vocation, he did have a hidden streak of play. Sometimes, he would make animal figurines for children from the glass pipette tubes in his laboratory. Other times, Fleming painted pictures in glass petri dishes using bacteria for his pigments. In addition to being a Picasso of germs, Fleming was also known among his peers to be a bit of a slob. While his colleagues cleaned and sterilized petri dishes soon after an experiment was done, Fleming let piles of them loiter on his bench for weeks.

After returning from a six-week summer vacation in the countryside in September 1928, Fleming faced towers of petri dishes, which he started to clean, sterilize, and store. Fresh from his holiday, he noticed something in one of them. In one of the dishes, the bacteria of staphylococci was everywhere, except next to where a mold grew. The bacteria didn't care for this interloper. Having a contaminant such as a speck of dust or a spore of mold was a common nuisance in the laboratory. But in this case, Fleming wasn't annoyed. He gazed down at the dish for a long time and said, "That's funny."

After lassoing the mold and growing more of it, he identified it under the microscope as penicillin and placed it in bacterial brawls with bad germs under glass. He observed that penicillin knocked out streptococcus, staphylococcus, gonorrhea, and meningitis, but it could not beat typhoid fever or dysentery. Penicillin seemed powerful, but to make penicillin useful for humans, much, much more was needed, which was beyond Fleming's temperament or training.

Fleming wrote up his findings in a science paper in 1929 hoping his work, like a message in a bottle, would reach an agreeable shore. Nearly ten years later, in 1938, Fleming's paper was noticed by Oxford researcher Ernst Chain. Chain, along with his boss Howard Florey and colleague Norman Heatley, turned their laboratory into a penicillin factory, to bring massive amounts of this elixir to the world's stage. Penicillin saved millions and millions of lives. But all this would not have happened if the strange speck of dust in a glass dish had not been noticed.

Glass is an ancient material with opposing traits. It can be strong like a car windshield or fragile like a Christmas ornament. What is clear about glass is that it is civilization's old friend. The Egyptians used glass to make beautiful vessels and ornaments that required a high level of skill. Today, optical fibers made of glass carry information from the internet. Glass, which originates from beach sand, has touched humankind in most aspects of life. With it, we've adorned churches, shielded light bulbs, covered skyscrapers, and even seen our reflection.

Glass has also played a significant role in discoveries. Glass has been used to see other worlds bigger than ours with telescopes, and smaller than ours with microscopes. "Seeing is believing" is the crux of science; glass is at the heart of this scientific method.

Today, no modern laboratory would be complete without test tubes, beakers, graduated cylinders, and flasks, dutifully waiting for action. With them, scientists and researchers have found the origins of—and cures for—anthrax, tuberculosis, malaria, and even Montezuma's revenge. As important as glass has been for science, we have looked through it, but we rarely look *at* it. Glass, itself, is rarely under the microscope. If the focus is adjusted onto glass, though, just as with many discoveries made possible with it, something new will be uncovered.

Through the Glass, Darkly

Otto Schott had dreams of discovering new things in the tidy and clean space of a chemical laboratory. Unfortunately, he was born into a family of glassmakers in Witten, Germany, in 1851, who worked in the heat, sweat, and dust of workshops. For generations, both sides of his family toiled in this strenuous and stagnant trade, and there was the expectation, spoken and unspoken, that he would join his father in the windowpane factory. Young Otto Schott had other plans, however. He took every chemistry class he could, starting from high school, to prepare for a doctorate in organic chemistry. Schott, a short and slight man with a handlebar mustache, wanted to leave his mark using his brain to understand materials, not his brawn to fashion them. Chemistry in the 1870s in Germany was the route to many exciting innovations, particularly in the making of drugs, fertilizers, and explosives. Organic chemists were enchanted with the ability to copy substances made in nature, like the flavor of vanilla, and artificially duplicate them in the laboratory. Nature did not give up her secrets easily, but when her methods were decoded, these molecules became new products, manufactured by the ton. One chemical victory that certainly piqued Schott's interest was the 1856 creation of a purple dye, called mauve, when William Perkins converted coal tar into the color that became the fashion rage. When Schott was a young boy, textiles had a limited palette of blacks, reds, and blues, all derived from plants, minerals, and animals. But mauve, which was cooked up in a lab, made it possible to create other bolder colors, by combining it with other pigments, and without the demise of living things. Germany became the biggest producer of this dye, having the monopoly on it, providing plenty of Perkins' Purple, as Charles Dickens called it, for the public's pleasure. The world was smitten with what organic chemists could do, including Otto Schott.

With visions of molecules dancing in his head, Schott applied for a graduate research position at the University of Leipzig for his doctoral work in organic chemistry. There was no space for him, though. Disappointed but undeterred, he tried to enter organic chemistry sideways, by taking graduate classes in agricultural chemistry. But he soon found this new topic uninteresting and dropped out. With his dream thwarted, he returned to glass, but this time for his doctoral studies, completing them in 1875 at the University of Jena, a popular and vibrant university, which Karl Marx once attended. The title of Schott's dissertation was "Contributions to the Theory and Practice of Glass Fabrication," a subject he knew well from the time he was a boy. After his graduate studies, he went on to work in a glass factory, publishing papers on the melting of glass, the strengthening of glass, and the chemical elements in glass. Schott returned to his hometown of Witten in 1878, steadily experimenting in glass on the factory floor. His work did not set the world on fire, but using fire and chemical ingredients he hoped to unlock the workings of this old material and make it new.

About 250 miles west from a wistful Otto Schott was frustrated Professor Ernest Abbe in a laboratory in the college town of Jena. An esteemed professor of physics and the director of a telescope observatory, Professor Abbe had grown leery of the glass lenses in his microscopes and telescopes. The professor, with his mathematician's unkemptness of finger-combed hair, a scruffy graying beard, and spectacles perched at the end of his nose, noticed there were multitudes of flaws in his scientific lenses, making it difficult to see anything clearly. Sometimes the glass had bubbles, streaks, or lines—called straie—that looked like a ship's narrow wake. Sometimes the glass was cloudy, hazy, or swirled, with unblended parts like a marble cake. Above all, the glass itself was inferior because colors, like blue and red, separated within an image, as if seen through modern 3D glasses. With such horrible materials, the chance of scientific breakthroughs was nearly impossible, since glass was at the heart of these instruments. Without good glass, science was blind.

To vent his frustration about the lack of research on glass, the good Professor Abbe did what any man of science would do. He wrote a report in 1876 stating that the future of the tweed-wearing scientists' fine optical instruments like microscopes and telescopes were in the thick and calloused hands of apron-wearing glassmakers. The earliest glasses entailed heating and mixing the ingredients of sodium carbonate (or soda), limestone (chalk), and silica (sand), creating crown glass, which was used for windowpanes and bottles. Replacing the chalk with lead compounds created the more ornate flint glass, also called lead crystal. These two families of glasses had been the only ones for centuries, and Abbe declared there was a scarcity of studies that explored new additives to make glass with improved optical properties.

In his report, Abbe set a new research direction, stating that "the development of new types of optical glass with uniform, calculable, and predictable properties was required." Abbe wanted the manner of how a glass interacts with light to be cooked into it. Just as a baker changes the amount of flour, water, yeast, and baking soda to modify a bread's texture and chewiness, Abbe wanted to know how the chemical ingredients changed the glass's ability to fan out white light into colors of the rainbow, or its ability to bend light, the way a straw seems broken in a beverage. Professor Abbe wanted these traits to be turned up or down in a regular and repeatable way, with the knob being the chemical elements that make up the glass. His report went on to say how little had been done in the study of glass for decades, and he stated openly what many knew but were too polite to mention, particularly that the making of glass was based on traditional recipes and not on technical knowhow. And without that knowhow, science was going nowhere.

Otto Schott got hold of this report three years later, and wrote a letter to the professor in 1879, volunteering to supply a range of different types of glasses, with the hopes of escaping the sweltering heat and silt of the factory floor. Schott had been working on systematically creating glasses of different chemical ingredients and of differing amounts, but he didn't have access to a laboratory to make the scientific measurements to see what his glasses could do. Abbe had access to those instruments, but didn't have the ability to make new glasses. Together, these men were each other's yin and yang. Professor Abbe was willing to collaborate with a person unknown in the sciences, because he had nothing to lose. Otto was keen on doing extra work, because he had everything to gain. This opportunity was Otto Schott's shot.

Schott sent glass samples to Abbe, but they did not have the optical properties that Abbe desired. Nevertheless, the two struck up a correspondence for over a year and a half and Schott continued to make glasses of different combinations of chemicals and of different amounts. Schott was able to make better choices than scientists of the past because twenty years earlier a Siberian scientist, Dmitri Mendeleev, turned the field of chemistry upside down with his breakthrough of the Periodic Table, where all of the known ingredients of the world—the elements—were found to systematically relate to each other on a chart, where elements near each other on this chart acted like cousins. Using the new periodic table, Schott began to take a methodical approach to explore how different concoctions of glasses behaved. New formulations under the guidance of this table allowed Schott to make better educated guesses.

Schott set plans in 1880 to make new glass mixes, treating the periodic table like a restaurant menu, selecting options from different columns, and sometimes from the same column, to see what worked best. He began his efforts by adding the elements of phosphorus and boron. In the fall of 1881 he focused on boron, which comes from borax (a detergent additive), and discovered something very promising. The adding of its boric acid into glass made a new type of glass, a borosilicate glass, which seemed free from defects. Schott sent these new glasses to Professor Abbe for testing, looking forward to the results. One day, Schott got a letter from Abbe congratulating him. In a letter dated October 7, 1881, Abbe wrote, "the problem," as Abbe wrote about the flaws of optical glass, "has been solved." The professor followed up on that note by inviting Schott to visit Jena to demonstrate his new glass.

With continual improvements over the next year, Otto Schott got his secret wish. Abbe wrote that he thought it best for Schott not to continue his work in a glass factory, but insisted that it be done in a chemical laboratory in Jena. Schott made arrangements to leave.

In 1882, Otto Schott moved to the city of Jena to run a small-scale operation, in partnership with Professor Abbe and Carl Zeiss (a microscope maker), with whom Abbe had a longstanding business relationship. No longer were Schott's experiments confined to small furnaces where he could create samples no bigger than a cup of sugar. Schott's specimens were now huge icicles the diameter of bowling balls. Schott started a company in 1884, Schott & Associates Glass Technology Laboratory, to make and sell specialized glass. Their first catalog, published by the company in 1886, had forty-four different glasses; by 1892, there were seventy-six.

Schott devised new formulations for better optical lenses, and later for thermometers, too. In the late 1800s, thermometers were one of the few tools that scientists had to probe a chemical reaction. At the time, chemistry was limited to knowing how hot things got (the temperature), how much they weighed (the mass), how much space they took up (the volume), and how much they pushed the container walls (the pressure). Many scientists noticed that their thermometer readings were higher than they should be. It turned out that thermometers did not return to the proper baseline once they were cooled. The heating and cooling and heating and cooling and heating and cooling that the thermometers underwent modified the glass so that the bulb, which held the mercury, changed its shape, causing the mercury to creep up. This meant that subsequent temperature readings were not to be trusted. By tailoring the amount of boron, Schott was able to build a glass that did not adjust its physique when heated, allowing thermometers to make proper readings.

Otto Schott cranked out different flavors of glasses in collaboration with Abbe. One was glass that did not alter its form with heat, allowing thermometers to display proper readings. Another was optically superior and perfect for scientific telescopes and microscopes. A final one did not dissolve in water, acid, or other liquids, making it suitable for laboratory experiments. At the heart of his new creations was boron, though boron played different roles in each of Schott's new glasses. Schott's glasses had versions with small, medium, and large amounts of boron, the way a chef can make a sauce with different levels of spice—mild, medium, or hot—based on the amount of pepper in it. For glasses with improved optical properties, a small bit of boron was added to windowpane glass, giving the glass a better ability to bend light. For glasses that did not expand when heated, Schott added lots of boron. Boron tightly grips other atoms with stiff bonds, like a strong spring, causing the resulting glass to resist expanding when it got hot, unlike other glasses. Finally, for glasses to be able to withstand dangerous chemicals like acids, the amount of boron was decreased to a medium level. Boron likes to bond with other atoms, but its bonds are weak in acids. So some of the boron was taken out of harm's way and substituted with other compounds. Together, all these ingredients stabilized the glass in these harsh environments.

Soon, the glasses designed by Schott became the most desired scientific glass on the planet, and Germany became the main source of all glasses for microscopes, telescopes, and laboratory ware (beakers, flasks, and test tubes). Every scientist wanted optical instruments with the name JENA inscribed on them. For other glassmakers, penetrating this glass market seemed impossible. One company in upstate New York realized that its only chance was to use science.

In the early 1900s, American glassmakers wanted to develop an alternative to Germany's Jena glass. Cracking the code for Jena borosilicate glass wasn't easy, though. Glassmakers knew that boron was a key ingredient, but the rest of the recipe was a mystery. Otto Schott spelled out in his highly technical papers the factors that allowed glass to withstand high heat and large temperature differences, but few glass workers could translate the theory of Schott's papers into practice on the factory floor. To be successful, one American company, Corning Glass Works of Corning, New York, knew that their apron-wearing workmen were going to need some help from academics.

Corning Glass Works was a family-run company that moved from Brooklyn, New York, to Corning in 1868 with the promise of the canal to transport products and coal from Pennsylvania to keep its furnaces hot. Corning mostly made decorative glass and tableware, and soon also hand-blown glass for Edison's light bulbs. If they were going to compete with Jena glass, though, they knew they needed more science to create new offerings. Corning started to move away from using glass recipes passed down from one generation to the next and began to apply the scientific method. One of the first things Corning's management did was tell their workers to write down what they added to a glass melt, so that a batch could be repeated if need be. Corning also began an unusual practice for a glass factory at the time: they hired scientists.

Starting in 1908, chemists were on Corning's payroll and the investment was working out to be a wise one. To differentiate themselves from other glass companies and to compete with German products, Corning needed technical personnel. Scientists at Corning knew that boron was a key ingredient in these new glasses, and eventually, by trial and error, they created a version of a borosilicate glass called Nonex (short for NON-EXpanding glass). Unfortunately, Corning could not penetrate the labware market with it. Its early glass was no rival to Jena glass, which had a nearly fifteen-year head-start. Additionally, German glass enjoyed low tariffs, since it was an educational product. Customers saw no reason to buy an American-made glass, when the price of the superior glasses from Germany was not prohibitive. Corning's management had to find a domestic market for their borosilicate glass and reached out to the most lucrative business in the nation to help keep the company afloat: Corning tapped into the big industry of the railroad.

In the early 1900s, the tentacles of railroads reached some of the farthest corners of the country. In addition to annihilating space, the railroad also compressed time with its speed. That speed came at a cost, however. As trains became faster, many more catastrophic accidents and collisions occurred and with that came a need for better signaling to increase safety. Signals on the tracks told trains not to proceed, with warnings from hot arc lights with red glass covers. On rainy or snowy days, however, accidents occurred more frequently. In addition to the inclement weather, another cause for the increase in accidents was the frailty of glass.

Glasses on train signals were between a rock and a hard place on days with bad weather. The interior of the glass signal was heated by the hot arc light, causing it to expand; the outside, however, was dramatically cooled by the rain or snow, causing it to shrink. The conflicting messages within the glass gave rise to pent up stress; when prolonged this stress resulted in broken glass. A red glass alerts the train to stop, but a broken glass is no longer red and tells the conductor to proceed, giving a false—and possibly deadly—message that it is safe to pass, potentially causing a colossal collision. And as if the weather was not enough for the glass to contend with, mischievous boys used train signals for BB gun target practice, smashing the red glass into pieces with a single pellet. The railroad needed better glass to mitigate the weather and the delinquents and Corning's strong Nonex glass did the trick.

Corning's glass rarely failed. Corning soon became a victim of its own success, however. When the railroad adopted Corning's glass, there was a boom in sales, but the indestructibility of the glass meant that once the railroads purchased their hardy glass, they didn't need replacements. The meteoric increase in sales was followed by a precipitous drop. This lack of built-in obsolescence, or a limitation that would have required additional sales, caused the company to scramble for new glass markets. Help would come, of all places, from a cake.

One summer afternoon in 1913, Jesse Talbot Littleton, a physicist and one of Corning Glass Works's newest scientists, came in to work with a sponge cake baked by his wife, Bessie. J. T., as Jesse preferred to be called, and Bessie were southerners; he was from Alabama and she was from Mississippi. They had moved to Corning from Ann Arbor, Michigan, where J. T. was a physics professor a year prior and, together, they were trying to get accustomed to their new Yankee home in Corning, New York. In a spirit of southern hospitality, J. T. Littleton brought in a cake. The cake was not just a social offering, though; it was a science experiment. For the last two weeks, J. T. had been trying to convince his colleagues of the benefits of cooking with a glass container, but his colleagues laughed at the notion. For generations, people had been told to keep glass away from heat. So baking with glass seemed ridiculous. Little did they know that Littleton was not only a southern boy, but also a glass man.

Littleton was obsessed with glass. He talked about glass at the dinner table. When Jell-O came for dessert, he'd break it apart slowly and show his children how it broke like glass. He even had hopes of being buried in a glass coffin. What made him so certain about his claims about the ability of glass to be a cooking container was that he did his 1911 dissertation at the University of Wisconsin on the heating properties of glass. For the rest of the scientists, who were chemists, the heating of glass was unknown. These scientists assumed that the thick walls of the glass would prevent food from cooking evenly and that the heat would not spread as well as it would with a thin metal pan. J. T. Littleton, a physicist, knew otherwise. When his colleagues did not listen to his words, his southern sensibilities didn't take kindly to being mocked. So he decided to follow them up with action. He got some help from Bessie.

Bessie Littleton liked company. She was raised on a remote Mississippi plantation where visitors were few. At her new home in upstate New York she asked J. T. to bring people from work over for dinner. At barely five feet tall, with her dark hair in a bouffant style, she was slight, talkative, and dogmatic. With Bessie things had to be just so and she had very strict rules that J. T. had to follow: no lying or liquor; no cigarettes or cigars; no cussin' or colored people at her table. The long and lanky J. T., with his tall frame, eyeglasses, serious eyes, permanent pout, and understated grace, complied by bringing over a fellow scientist, H. Phelps Gage. All evening, Bessie chided Gage, who was a bachelor, to get married. While the men talked about glass after dinner, Bessie had a captive audience for something that had been troubling her.

A few days earlier, Bessie's new Guernsey casserole dish, which she only used one other time, broke. All night the men talked about the indestructibility of glass and she insisted that these smart alecks ought to make cookware that did not break. The next day J. T. got two cylindrical Nonex battery jars, about as wide as a basketball, cut off the bottom to make round dishes, and brought them home to Bessie.

Bessie didn't cook. She had servants to do that. As a child in the south, her servants were freed black slaves who could not escape the grip of the plantation. As an adult in the north, she hired white immigrant girls, whose families came to New York State for work.

While Bessie was no master chef, she dominated with her baking. As soon as J. T. gave her an indestructible glass dish, she immediately got to her favorite kitchen task and turned sugar, eggs, flour, butter, milk, vanilla, and baking powder into a white cake. Using every bowl and utensil in the kitchen, she poured the batter into her newfangled dish and baked. What emerged from the oven was an evenly brown cake with a color surpassing that provided by her metal dishes.

The next day, J. T. Littleton brought the cake into work and everyone, not knowing about the baking experiment, reported that the cake was good. Littleton then told them that this cake was baked in glass, causing scientists to scratch their heads and men in management to rub their chins.

These scientists found that the cake actually baked well with an inviting color of brown on top. Littleton relayed to his colleagues how easy it was to remove the cake from the smooth glass pan, unlike metal cake pans. His science colleagues did not think glass, if it survived, could make a cake as delectable as the one Jesse brought in, and, in a sense, they ate their words.

They asked that Bessie try out other foods and report how the glass pan worked. So Bessie, as the resident domestic scientist, had a few items cooked, from French fries to steak to cocoa, although she had a penchant for southern dishes of grits, cornbread, and collard greens. The pan performed well, the food didn't stick, and the glass pan didn't retain the flavor of the food the way a metal skillet did.

On hearing about the success of this glass for cooking food, the Corning management saw promise. But they had to make a few changes and learn a few more things. First, the formulation of Nonex had to be altered, since it contained lead. The scientists made a borosilicate glass without lead for this bakeware. Next, they had to test the strength of the glass, dropping a weight as heavy as a can of soup onto different types of dishes to see how they survived the rigors of a kitchen. While earthenware cracked with a weight dropped at six inches and crockery broke at ten inches, borosilicate glass laughed off the impact, untouched, even when the weight was dropped waist high. After these impact tests, the team had to figure out how the glass cooked food. Bessie reported that food cooked quicker than in a metal pan, which was the opposite of what they believed would happen. They got to the bottom of this with an experiment.

A scientist dipped a Nonex pan into a liquid chemical bath full of microscopic bits of silver. The silver settled on the surface, coating the outside with a thin layer, giving it a mirror finish. Then, they baked two cakes: one in a simple Nonex pan and one in the mirrored one. After baking, they noticed that the cake with the silver coating did not cook well. What they learned is that the heat from the oven walls, like the rays of the sun, transmitted through the clear glass, cooking the cake, while the mirrored surface reflected that heat back. This showed them that the glass pan cooked differently than a metal pan. A cake in a metal pan gets heated from the hot air in the oven and the heat from the oven rack. The glass, meanwhile, was letting heat into the cake a third way, from the invisible rays of heat, like the sun, which browns our skins and the crust of a loaf of bread.

To commercialize this glass with a new purpose, it needed a name that informed the consumers, mostly women, what this new glass did. The first commercial piece on the market was a pie tin, which was initially called "Py-right." It was renamed Pyrex in 1915, to relate to the earlier product, Nonex, and to sound more futuristic and medical, like Latex or Cutex. Sales of Pyrex were initially flat, but after the company listened and responded to customers' needs, for example reducing the weight of the ovenware, Pyrex soon became a standard item in households. By 1919, over 4.5 million pieces of ovenware were sold. To encourage more sales, Corning created many shapes and sizes and colors, learning their lesson from the railroad glass episode, which made Pyrex a standard Christmas gift. Corning still had its eye on glass for labware, however. An opportunity to enter that market would be a gift of war.

In 1915, with America's potential entry into the war, it became clear to the American government that it needed the ability to make glass for military applications. Jena glass was regarded as the best in the world, but imports from Germany were dwindling. American companies, such as Corning Glass Works, were encouraged years before to create a German glass substitute. Legend has it that President Woodrow Wilson asked Corning's management to develop an alternative to the German products in preparation. These glasses would be used by American soldiers in gun sights and binoculars, by sailors in sextants and periscopes, by airmen for aerial cameras and rangefinders, by army doctors for thermometers and vials of medicines, and in the laboratory by chemists for the synthesis of explosives.

At the precipice of America's entry into the war, Corning had a borosilicate glass, although the ideal Jena formulas were still locked up in German patents. Corning and many other companies wished they could get hold of these recipes. They would get their wish.

What the American companies may not have known is that laws of peacetime do not hold during war. When the United States entered the war, it confiscated thousands of German patents (nearly 20,000) as part of its war booty. Impenetrable German monopolies protected by patents, for dyes like mauve and drugs like aspirin, were blasted open with one of America's secret weapons. This weapon was based not on combustion, but on the legislation of the Trading with the Enemy Act. With it, German science, the science of the enemy, became fair game to Americans and American companies. Buried within those patents were those recipes for specialty glasses.

After the war, Corning introduced a range of new Pyrex products, filling in for the shrinking supplies from Germany. In laboratories, now there were Pyrex petri dishes, test tubes, and flasks. In homes, there were Pyrex cooking dishes, oven door windows, and percolator tops. In automobiles, there were Pyrex headlights, battery jars, and pressure gage covers. America had unknowingly entered the Glass Age, whereby Corning created a new American industry of scientific and specialty glass. To keep this comfy cushion of no competition for their consumer commodities, Corning pushed for legislation, a tool they grew to understand, to prevent the influx of German glass into American markets after the war. Huge tariffs were placed on German glass, preventing Germany from monopolizing these markets as they had done in the past.

These actions were out of view for most Americans and most scientists, who used Pyrex glasses to find the causes for diseases in glass petri dishes, and developed drugs to fight them in glass test tubes. What citizens and scientists did not know was that glass also provided containers that cooked up a new narrative of American innovation and its scientific prowess. There was no doubt that America was a science superpower, but what was unknown at the time was that the United States now had the upper hand, particularly in glass, made possible by a curious combination of war and cake.

No scientific lab was complete without glass. Through the use of glass, we gained an understanding of how our bodies work, how the heavens move, and how other worlds exist in a drop of water. Glass helped to change our perspective.

Ironically, glass helped to order our lives, but its transparency is brought into being by the chaos within it. The atoms in glass are not given enough time to line up like soldiers, so they sit frozen in place in disarray, as in a snapshot of kindergartners during recess. Glass is full of disorder, but the clarity of glass helped us make sense of the world with the lenses and beakers and flasks it created. Since antiquity, glass was treasured for its beauty, but glass also allowed for the cooking up of new drugs, formulas, and medicines. At the end of the nineteenth century, glass also helped a scientist who was no friend to glass to discover the future.

J. J.'s Ray Gun

Long before the Great War, in 1895, science and magic were hard to separate. That year, Wilhelm Roentgen took a ghostly picture of his wife's hand using mysterious rays that showed her bones. These invisible rays, later called X-rays, shot out of a contraption made of metal and glass that looked like something out of Dr. Franken-stein's laboratory. Newspapers packed their pages with depictions of a person's insides on the outside, and readers snatched up copies. Scientists were also enchanted by X-rays. Some of them wanted to know what else they could do. Others wondered where they came from. All these scientists understood that a battery attached to a stretched glass globe spawned a glowing stream called a cathode ray, and when this cathode ray collided with a piece of metal inside the globe, out came X-rays. Their thinking was that there must be more to these cathode rays. So while the whole world was wowed with X-rays, a few scientists were hoping to find the next great thing in cathode rays. Little did they know that this bright stream would explain how the world worked.

Cathode rays had been known for decades, but there was little consensus about their origin, and eventually the case went cold. With the renewed interest in them, scientists obsessed over every move of cathode rays, writing articles with reports of their behav-ior, though not yet knowing that cathode rays held a key to their scientific understanding. Locked in those cathode rays was the cur-rency of all chemical reactions. Locked in those cathode rays was the answer to science questions from how toasters work to how planets were born. Locked in those cathode rays were the droplets that powered a river of modern technologies from televisions to computers to cellphones. Unbeknownst to these early scientists was that inside the cathode ray was a part of the atom that they didn't know existed—the electron. But deciphering the puzzle of cathode rays required uncovering clues. Just as the popular character Sherlock Holmes used his intellect and his magnifying glass to solve mysteries, scientists too had to observe cathode rays under glass. For some scientists, this puzzle was too delicious to turn down, and Joseph John Thomson was one of them. It was this short man from the nineteenth century who would make the giant leap that made the technologies of the twentieth and twenty-first centuries possible.

Thomson's potential in answering one of the biggest questions of his day seemed doubtful when he was fourteen years old in 1870. All he wanted to be was a botanist. As a small boy growing up near the city of Manchester, England, he spent all his pocket money on weekly gardening magazines. His father, a modest bookseller, wanted him to have a stable trade as an engineer. Being an engineer was good work, as Manchester's textile mills turned American cotton into goods. To please his father, J. J., as Joseph John Thomson was nicknamed, attended Owen's College in Manchester in 1870. But when his father died, J. J. scrambled to stay in school by winning scholarships. He entered Trinity College in Cambridge to study mathematics, choosing the beauty of numbers, instead of their utility, as in engineering. Walking on the hallowed grounds that Sir Isaac Newton strolled was an achievement for any son of a bookseller. But J. J. never fit in.

J. J. may not have felt at home at this old university, but his genius certainly was at home there. By 1895, Thomson was the thirty-nine-year-old head of the Cavendish Laboratory at Cambridge University, blossoming into an absentminded mathematics professor. His eyeglasses had two positions—one on his nose, which meant he was thinking, and the other on his forehead, which meant he was thinking more. He did not trouble his brain with worry about his appearance so his hair was long, his mustache overgrown, and his chin badly shaven. His brain was congested with abstract ideas, so his new research on cathode rays meant there'd be even less space to worry about ordinary things.

Uncovering the origin of the cathode rays was a perfect puzzle for J. J. because it challenged him by linking abstract ideas with observable events. Cathode rays shot from one electrical connection to another inside of a glass tube without air, and there were two dueling beliefs among scientists about how cathode rays moved in the world. One group thought that cathode rays were a wave that was a wrinkle in the ether. Others concluded that the beam was made up of small bits of particles acting together, like a migrating flock of birds. "Neither side was wholly right nor wholly wrong," said J. J. There was evidence to support both ideas, but the cathode ray could not be both.

One definitive way to see if a cathode ray was a wave or a particle was to observe its dance with magnets. There was an old theory that said that if cathode rays fly undisturbed by a magnet, they are a wave; and if a magnet deflects the ray, they are made up of particles. J. J. wanted to test this theory and learned that a few years earlier, in 1883, another scientist performed this very same experiment. Cathode rays did not move when a magnet was nearby, supporting the wave argument. But J. J. thought there was something wrong with that earlier attempt. Scientific tools had advanced since then, and could draw more air out of a glass tube to better create a vacuum. A vacuum with less air was the habitat where cathode rays thrived best. So J. J., who believed that cathode rays were full of particles, wanted to repeat this old experiment using a glass tube with less air in it, made possible with an improved vacuum.

J. J.'s mathematical genius, unfortunately, did not translate into manual dexterity. For such a small man, he was a Victorian bull in a china shop. When he visited his students in the laboratory, they'd wince when he offered help, and quickly tried to move fragile things out of his way. They took deep breaths when he sat on a lab stool to speak. Life was no better at home. J. J.'s wife did not permit him to use a hammer in the house.

J. J. needed help with his experiments and that help came from a former chemistry assistant, Ebeneezer Everett. While the name Ebeneezer conjures a miserly image, Everett was a dashing, mustached man, with cowboy good looks, who leaned a bit to seem less tall. Little is known of this Everett, except that he was a patient soul and a virtuoso for making laboratory glassware out of common soda lime glass into works of art that would have pleased a Murano glass master. Lab benches were full of Everett's glass constructions, braced in place with wood brackets, with wires on every surface and sticking up into the air. Everett was the scientific brawn to J. J.'s brain.

Starting in late 1896, J. J. wanted to make a cathode-ray obstacle course to settle this wave/particle debate. Everett made a sophisticated glass bulb with pieces inside, reminiscent of a model ship in a bottle. On one end of the glass two metal pins stuck out that were attached to the ends of a battery to produce the cathode ray. Inside the glass, the cathode rays sprayed out in many directions like water out of a hose and were focused into a narrow stream, with two slits that acted like a nozzle. That beam then hit the interior surface of a round bulb, creating a green glow.

Cathode rays required that there be very little air inside the glass tube. "This was more easily said than done," said J. J. To remove the air, Everett poured liquid mercury into a tower, which he connected to his glass bulb with a glass bridge. As the heavy liquid fell, it sucked air across the bridge from the glass bulb, creating a vacuum. Removing the air sometimes took most of the day, so Everett started in the morning before the hurricane in the form of J. J. Thomson arrived in the laboratory in the afternoon.

Only glass worked for these experiments. Copper would not do, nor any metal for that matter, for metals would bury the cathode ray. Wood or clay would not work either, for they could not hold a vacuum. Clear plastics hadn't been invented yet. Glass was the best keeper of a vacuum; transparent, uninterested in conducting electricity, and malleable to an inventor's imagination. But, mostly, glass was vital in science because it allowed scientists to do what they do best, which is to use their power of observation—and this was what J. J. excelled at.

Sometimes J. J. complained to his colleagues about his glassware. "I believed all the glass in the place is bewitched," he said. Standard recipes did not yet exist for glass. Some parts of a glass tube were richer in key ingredients than others. To build with glass required compositions that were uniform all over, so that they would melt at the same temperature. And a glass piece divulged how well the bond was made only after many hours of work had passed. Sometimes glass whispered with a small air leak that there was something wrong, other times it screamed with explosions. Glass was temperamental, and it was up to Everett to tend to it like a newborn baby.

In the summer of 1897, Everett completed J. J. Thomson's obstacle course for testing cathode rays. He inserted two additional metal plates and attached them to another battery, creating an electric field, as a way to nudge the rays. As Everett turned the contraption on, J. J. saw that the cathode ray moved downward to the metal plate connected to the positive end of the battery. This told J. J. that the cathode ray was negative. Everett then put a huge horseshoe magnet around the center of the glass tube, and when he turned it on, J. J. saw that the cathode ray moved up, like migrating birds swept up by a strong wind. From J. J.'s mathematical calculations, written on the backs of random scraps of paper, he was able to deduce that the cathode ray was made of small bits that were electrically charged and negative. He calculated they were smaller than an atom, and were thus the tiniest part of matter yet discovered.

And when he and Everett repeated these experiments with different metal plates and with different gases inside the tube, J. J. saw that these same small negative charges existed in all materials. He called these bits corpuscles, but they would later be known as electrons.

J. J.'s discovery changed the world, but he could not predict that it would. This small and odd man found the small and odd electron, opening up a door in science and expanding the understanding of matter. The discovery of the electron gave us clues about how galaxies and stars and atoms formed, and the exchange of electrons between atoms, in chemical bonds, explained how hot gases from the Big Bang eventually became us. This discovery also revealed the basic building block of technology. With the electron, scientists would come to understand the workings of circuits, static electricity, batteries, piezoelectricity, magnets, generators, and transistors. With the knowledge of electrons, technology—and society—blossomed.

When J. J. Thomson was growing up, many inventions that we now take for granted did not exist. There was "no car, no airplane, no electric light, no telephone, no radio." But the electrons in his glass, which made up electricity, would power all these machines as well as later developments such as computers, cellphones, and the internet. As smart as J. J. was, he could never have predicted that this abstract science would have practical implications. But it did, and it had many. With his discovery, humanity was thrust into a new age—an electronic one. None of these technologies, however, would have happened if it weren't for the ability to see electrons in action. Our modern world was made possible by the ancient and old material of glass.

8
Think

How the creation of rudimentary telephone switches ushered in silicon chips for computers, but also rewired our brains.

Googled

Phineas Gage should have been dead. On the afternoon of September 13, 1848, a terrible accident occurred on an ordinary Wednesday at a construction site not far from Vermont's Green Mountains. Gage, a handsome twenty-five-year-old railroad foreman, packed gunpowder into a hole, as he'd done hundreds of times to focus a blast, using the flat end of his "tamping" rod. But this fateful time, Gage wasn't paying attention. The rod, shaped like a mammoth sewing needle, was in his grip when it scraped a rock, creating a spark. The ignited gunpowder propelled the three-foot seven-inch bar into his face, sliding under his left cheek, behind his left eye, through his brain, and out just beyond his hairline, clanking twenty yards behind him. The thirteen pounds of iron, with its tip as wide as a pencil and a base the width of a silver dollar, rocketed upward, while Gage thudded to the ground. A few moments later, Gage's motionless body revived, like Lazarus, and he was soon giving an account of what happened, even helping himself onto a wagon to get medical attention, as blood poured from the holes in his head and face.

Gage would live another eleven years; his doctor said he had an "iron will as well as an iron frame." Although his body was more or less intact, his mind was not the same. Before the accident, the dark-haired and tall Gage was an affable, reliable, and clever young man who was a favorite among his workers; after the accident, he was irascible, erratic, and childlike, and he cursed excessively. Many of his friends said that after the accident Gage was "no longer Gage." His Jekyll and Hyde metamorphosis showed early physicians how the brain can be altered. Today, neuroscientists have discovered more about the brain and have learned that it can be modified in big ways and small. The brain is actually changed by its environment. For Gage, a bar of iron noticeably and instantly altered his personality; for us, our brains are slowly and stealthily being transformed by our computers and the internet.

While the brain is still a mystery, we understand more about how it works since the time country doctors examined Gage. Scientists know that certain regions of the brain have specific functions. Gage's brain was injured in the very front part of his head, and there lies the clue as to why his behavior was altered.

The shape of the brain can be likened to half of a grape on a stick, with a sprig of garnish on the bottom back-side. The grape is the cerebrum; the stick is the brain stem; and the garnish is the cerebellum. The brain stem regulates automatic functions (like breathing and the heart beating) and the cerebellum controls balance and coordination, but the cerebrum makes us *us*. The cerebrum is where we think, feel, remember, speak, create, and sense. The front of the brain, called the frontal lobe, controls executive functions like attention, focus, organization, and impulse control; it was this part of Gage's brain that was run through by the iron rod, explaining why he was so distracted, unreliable, fitful, and spoke like he had no religion. The front of Gage's brain was the concern then, but the parts of the brain that process and store information are a concern for us today, since their functioning is being altered by our devices.

For generations, a belief was held that after a certain phase of life the brain was hardwired for good. Scientists at the time were convinced that there was no way to create new connections, to learn new things, or to acquire new skills. The brain, in other words, was considered an old dog that couldn't learn new tricks. This meant that one could never learn to speak Spanish, or play the guitar, or learn to cook Southern cuisine when one got older. Scientists know differently now. The brain can learn new things; it is moldable, and it can be rewired. Scientists would call the brain plastic.

The molding of our brain has been part of our evolution. Our brains are originally *Homo sapiens* brains, created nearly 200,000 years ago, from descendants who originated from Africa. Our brains are Stone Age brains, but our technologies enhanced them. The simple tool of fire allowed ancient *Homo erectus* brains to grow. By reducing the energy required to chew and digest raw food, cooking freed up the body's resources, which helped to build big brains. Some time later, the printing press allowed us to share ideas with impressions of moveable type pressed on a page. By disseminating information widely, the book expanded our minds with greater opportunities to think more broadly. Our brain's adaption did not end there; it has continued even in the last century. The generation growing up listening to the radio has different auditory and imaginative skills than generations that grew up with the visual skills enhanced by television. The internet, and the computers that enable it, are the next technology to stretch our plastic brains.

The changing of our brains doesn't take much time, either. It happens within a lifetime. Scientists have probed—and have proven—the plasticity of our brains, using special cameras, employing a technology called magnetic resonance imaging (commonly known as an MRI), which can look into a living brain and watch it work. Researchers have found that skilled musicians have a part of the brain (in the cerebral cortex) that is bigger than that in nonmusicians. London taxicab drivers, learning streets by heart, increased the memory centers of their brains; even people who learn how to juggle in a several-week scientific study enlarged a part of the parietal lobe of their brains. These and many, many other studies have illustrated that we can modify our brains. This news is both wonderful and worrisome. The malleability of the brain is a gift, a superpower of flexibility of the three-pound marvel in our heads. But this capacity also means that our brains can be transformed with or without our doing. In our current time, the use of the internet is persistent, pervasive, and always present. This means that the web is not only expanding what we can do, but changing how our brains think.

There are many similarities between our brains and our computers. The brain is made up of complex pathways with methods for sending information to other parts, processing information, and storing information, too. The computer also has circuitry to send information through tiny metal wires; however, the processing of information and the speaking to other parts of the computer would require human brains to develop another ingredient, which would take a few centuries for computers to be as we know them. The maturing of modern computers hinged on the creation of the silicon transistor, which can start and stop the flow of electricity, like a faucet. As simple as the shutting off and on of current was, this ability was enough to make a language for computers based on these binary states of "on" and "off," which allowed transistors to speak to each other. Together, transistors along with their binary code made them greater than the sum of their parts. Inside computers, transistors send a message to other transistors to process, to calculate, or perform logical operations, enabling the whole computer to think. But all those parts would not come together until the twentieth century.

The path began in the nineteenth century, with the ancestor of the silicon transistor, a simple switch.

The silicon transistor inside our most elaborate computers spawned from a desire for an electrical switch. These transistors, which are today manufactured every year in astronomical numbers, led to the evolution of the computer, and eventually led to our brain's evolution. The dance between the human brain and the silicon brain made visible how the creator is re-created by its creation. But first, before the computer was even a thought, the humble desire to talk to one another across telephone wires gave a mortician-turned-inventor an idea that changed the course of humankind's next two centuries. This journey to the transistor began one Friday evening in New Haven, Connecticut, in 1877.

Teapot Handles and Undergarment Wires

On April 27, 1877, people lined up to pay the admission of 75 cents at the Skiff Opera House in New Haven, Connecticut to see a Friday evening "telephone concert" by thirty-year-old Alexander Graham Bell. Bell invented the telephone in 1875 and had made waves at the 1876 Centennial Exposition in Philadelphia, where Lord Kelvin was so enamored by the "wonderful thing" invented by Bell that he had to be pulled away. On the bare New England stage, Bell stood next to a small table with his telephone, consisting of a rectangular wooden box, long enough to hold a shoe, with a spout on the end. Another of these boxes was suspended from the ceiling, and one other was in the rear of the hall. Bell spoke into the spout, and a disembodied voice came out. Upon hearing the voice, the 300 people in the audience spontaneously responded with a thunder of applause.

The originator of the voice was Thomas A. Watson, Bell's twenty-three-year-old assistant, who a year earlier heard Bell beckon him, when Watson was in an adjacent room in a Boston attic. Now, separated by thirty miles with Watson located in Middletown, Connecticut, telegraph wires carried their voices to and from their telephones. The New Haven audience was enthralled as they eavesdropped on this telephone conversation. "A more interesting performance was never given in this city," stated the *New Haven Evening Register.* After the demonstration, Bell explained how the telephone employed vibrations. Bell also pontificated on how his gadget would be in every home, connected by a central office. That idea took root in one audience member; his name was George W. Coy.

Immediately after the lecture, Coy spoke with Professor Bell with the hopes of bringing telephones to Connecticut. Coy, with his walrus mustache making his boyish face look older, was a Civil War veteran who'd lost the use of his left hand, but that did not stop him from being industrious. He worked as a local manager at the Atlantic and Pacific Telegraph Company for eight years and seemed to be in it for the long haul. After learning about the telephone, however, Coy said, "I commenced at once," in the making of a telephone exchange system. Bell granted him the right to form a telephone franchise on November 3, 1877, and a few months later, Coy opened the New Haven District Telephone Company, where telephones, deemed "the greatest invention of recent years," could be connected with the switchboard Coy devised.

Coy obtained twenty-one subscribers of butchers, apothecaries, private homes, and carriage makers, enabling him to formally start his business on the cloudy and snowy winter day of January 28, 1878. His central office, located on the ground floor of the brick, six-story Boardman building in New Haven, was situated on a busy downtown corner at 219 Chapel Street. This narrow and short storefront office, shaped like a railroad car, was sparsely decorated with a crate on its side for a desk and a soapbox for a chair. Customers were offered the only real furniture of a well-worn armchair. Resting on a table, leaning against the wall, sat a doormat-sized, wood board, of about two by three feet. This was Coy's ticket to the future and the heart of his operation—the switchboard.

Grabbing materials readily available to him from his city and his home, Coy hammered carriage bolts into the black, walnut wood board, which were the endpoints of customer's telephone lines. Connecting the bolts were levers, with handles taken from teapots. On the backside, the bolts were attached to each other with wires that originated from Mrs. Coy's undergarments. Out from the board came thick, telegraph wires that went out the back window and onto roofs and treetops, and eventually into the homes of his customers.

Connecting two people on Coy's contraption required receiving a subscriber's call and moving its electrical signal across the switchboard with a series of switches, to the call's final destination, the way a pawn moves across a chessboard. Numerous steps were required to do this. First, the caller at home pressed a button ringing a bell at the central office, alerting the operator that their assistance was needed. The operator moved a switch to connect to their line. Click. To hear the patron with their headset, the operator moved another switch. Click. After learning whom the caller wanted to reach, the operator put the patron on hold, switching their headset "off." Click. The operator made a new connection with another line, to a second party, with another switch. Click. To ring that person, the operator connected a buzzer to that line with a switch. Click. The operator switched their headset back "on" and waited. Click. When the other person finally picked up, the operator switched their headset "off." Click. Then, the conversation, made possible by a cascade of clicks from switches, commenced.

As rudimentary as this switchboard was, able to handle only two conversations at once, it helped to achieve Alexander Graham Bell's prediction that telephone wires would come to the home, "just as gas and water are supplied," and that citizens would see the telephone as "not luxuries but necessaries." To make electrical telephone wires a possibility like a gas and water supply, though, a way to shut on and off electrical signals was needed. Water had faucets, gas had valves, and the electrical signals of phone calls now had switches, which were at the heart of Coy's invention.

Soon the telephone took hold of the public's imagination and the central office and the switchboard grew to match the demand. When switchboards began, young boys handled the work. At their zenith, switchboards and telephone exchanges were staffed by scores and scores of women, who were more polite and better-behaved than boys. The growth of the number of phones gave rise to a dance of more switches, more girls, more switches, more girls, followed by more switches. As the switchboard became simpler, the job of the operator became more complex, by troubleshooting problems and making decisions on the customer's behalf. In essence, the female operator, or "hello girl" as they were called, became the switch. In the end, the telephone ushered a link between telephone machinery and girls, and this symbiosis would remain until a temperamental mortician from Kansas City came along.

A Secret Undertaking

Almon Strowger slammed the door behind the telephone repairman. This was a weekly ritual for this hot-headed undertaker from Kansas City, Missouri. Beginning in 1888, Strowger developed the habit of calling, complaining, and cursing at the central office of the Missouri and Kansas Telephone Company, believing that his phone wasn't working. On a typical visit, a blameless repairman took the ten-minute walk to the undertaker's business, on West Ninth Street, to hunt for the cause of Strowger's problems. The repairman performed the ritual of testing the line, by turning the crank on the phone to ring the bell at the main switchboard. When the operator picked up, she successfully placed calls out from Strowger's line and then to it. All worked well and the repairman noted in his report that the trouble could not be found. Despite this, Strowger wasn't satisfied. He was convinced that he was losing business from lost calls and suspected that the "hello girls," or telephone operators, were the culprits. Strowger swore he'd do something about them.

Almon Brown Strowger (1839–1902) was a short man with an even shorter fuse. Born a few years after the birth of the telegraph, Strowger grew up in Penfield, New York, on the outskirts of Rochester, in a town where his grandparents had settled two generations before. Despite his deep familial roots in this city, Strowger suffered from wanderlust and on his 22nd birthday, enlisted in the Civil War, joining Company A of the New York 8th Calvary. At a mere 110 pounds, Strowger, with his full-beard and stern gaze, made up for any shortcomings with his scrappiness and fearlessness as a bugler on the front lines, guiding the actions of the troops with his sounds of charge. He was wounded at Winchester, became a second lieutenant, and was honorably discharged December 8, 1864. But the war never left him. He was known to be erratic, irascible, and just plain ornery.

After the Civil War, Strowger bounced around to different states, including Ohio, Illinois, and Kansas, teaching and farming in each. When he arrived in Topeka, Kansas in 1882, with his second wife and two girls, he decided that he wanted to be his own man with a distinguished vocation. Since training to be a doctor or dentist took time and money, Strowger took classes in undertaking.

In 1882, Strowger bought William McBratney's practice in North Topeka, finding the work to be steady. He wanted to grow his business, though, and needed to move to a more populated town. By 1887, he relocated to Kansas City, Missouri, and bought another undertaking business to do so. But soon his troubles with the phone company began.

One day, as Strowger came into his office, he took off his dark work overcoat and sat at his desk reading the newspaper. While perusing the obituary section, he saw that a friend of his had died and that the body had been prepared by a competitor. It isn't clear what troubled Strowger more, the loss of a friend or the loss of a customer as he entered into one of his epic tantrums. What is clear is that Strowger surmised that the telephone operator was steering business away from him.

Strowger flew into a creative fit. He jumped onto his desk chair, opened the desk drawer, and found a round box full of paper shirt collars, which he emptied into the trash so he could use the box. Next, he got long straight pins and pushed them into the side of the box, ten across and ten down. Imagining that these one hundred pins were connected to the wires of one hundred telephones, he then spun a pencil in the center, like a clock's minute hand, touching each head of the pin in a row. A pencil attached to a rod, moving up and down like an elevator, could reach every pin, where the motion was powered by a battery. He imagined that if a caller wanted to call telephone No. 67, a connection could be made by moving six steps up and seven over. He envisioned a step-by-step switch and believed with the right combination of magnets, motors, rods, and gears, the pencil could connect to a pin and place a call without the need of a human operator. The notion that the days of the "hello girls" were numbered lifted his spirits.

One day, while complaining about his phone to the central office, he reached a manager, Herman W. Ritterhoff, who paid him a visit and felt Strowger's wrath. Ritterhoff, a good-tempered man known for his hearty laugh, was able to calm Strowger down when he spotted the cause of his telephone troubles. The outside sign was touching the telephone's wires, short-circuiting the line, preventing calls from getting through. This observation made Strowger so elated that he felt compelled to share something with his new friend. Strowger showed Ritterhoff a crude diagram of his machine to replace the "hello girls." Ritterhoff saw it, didn't think much of it, laughed, and left.

Unbeknownst to Ritterhoff, Strowger was actually onto something, creating a machine to connect calls, or an automatic telephone exchange. It was the next step to fully realize telephone services on a massive scale with the creation of an automatic switch.

Strowger left Kansas City in 1891 to create a company, Strowger Automatic Telephone Exchange, in Chicago that made equipment based on his collar box, pin, and pencil design. On November 3, 1892, he inaugurated his first system in the city of La Porte, Indiana. Inside the exchange office were walls full of shelves from top to bottom with dozens of Strowger switches, tapping phone connections and clicking like woodpeckers.

In a customer's home, the wall-mounted phone had five levers, resembling diving boards. On the end of each board was a label: "0," "10," "100," "1000," and "R." A customer placed a call by pushing these levers. If a customer wanted to call a phone number, like 73, the "10" bar would be pushed seven times, and the "0" bar would be pushed three times. When a call ended, it would be released by pressing "R."

Back at the central office, a team of Strowger's devices would do its magic with the electrical impulses made by the customer's pushes. A wiper on a rod, the former pencil, moved vertically seven times, and then horizontally three times, following the customer's command. In Strowger's invention, electrically powered wipers moving up and down, clicking, and spinning around, made connections to telephone lines. As the number of customers grew, phone numbers lengthened, and telephone exchanges expanded. Strowger's switches, working together, received the call, *click*, sent the call to a section of town, *click*, then to a street, *click*, and finally to a home, *click*. This automatic switch was made to take the human out of the telephone network, but this invention would take the human out in other ways not intentionally designed.

In a generation, Coy put switches on telephone boards and Strowger automated them. Soon, it became clear to the management of telephone companies that there wouldn't be enough girls or reliable switches to handle the growth in customers. What was needed was a switch that was very small. Several decades after Strowger's invention, that answer came in 1947 with a contraption that looked like a science fair project gone wrong. This gadget was hideous, with a thin slab of a silvery stone, a plastic triangle and a gold ribbon, all held in place with a paper clip. To physicists, however, it was a thing of beauty: the transistor. It was a switch in miniature. But this switch was more than that. Transistors would later become the heart of modern computers, using a language of binary, and enable a machine to think.

Transistors controlled the movement of electricity. Without them, electricity was unbridled, like a mustang; with them, electricity could not only be steered, but put to work, like a mule. Strowger's switches and also vacuum tubes, looking like more complicated light bulbs, served the telephone company as switches. But Strowger's switches wore out and vacuum tubes broke easily, burned out, and used lots of power. The transistor was less fragile and needed less electricity to work. The discovery of the transistor trumpeted a new age—an electronic one. With transistors, big mechanical machines could become small; with transistors, less space held more circuitry. Every scientist wanted to be part of the transistor's making. One of them would have to take the long route, starting from Texas.

Gordon Teal

In 1930, when Gordon Teal arrived at Bell Labs with his doctorate in chemistry nearly complete, he was ready to be one of its shining stars. Unfortunately, everyone at Bell Labs wanted the same thing.

Teal found that Bell had a pecking order: the physicists orbited the stratosphere, writing theoretical formulas on chalkboards; the metallurgists flew above the treetops, applying practical knowledge on workbenches; and the chemists lived below the earth's surface, making what others dreamed up in glassware. Being a chemist was a supporting role, not the starring one. It seemed that the other chemists around him were perfectly happy with this arrangement, but just like the liquids boiling in his lab, Teal was bubbling over to do more.

Gordon Teal was the smartest boy the central plains of Texas had seen in a while. And, he knew it. Born in Dallas in 1907, he grew up loving all things scientific and his thin, wiry frame was never far from a mystery book. Teal personified the old adage that still waters run deep. He was one-part southern modesty and one-part silent ferociousness, all behind an expressionless face. He was a shy, soft-spoken Texan, whose mother, Azealia, implored that he be excellent and a good Baptist.

Teal went to school at Baylor, not far from home, and then headed east for graduate school at Brown, a Baptist school, as was his mother's preference. While at Brown, Teal's love affair with the element of germanium began. When he studied it, germanium was just a scientific curiosity, with very few practical uses, and Teal subjected it to various chemistry experiments and solutions. This pair of man and material were both understated and misunderstood. The man hid emotion and germanium hid a chemical trait. Teal was dedicated to germanium and brought his interest and expertise for all things pertaining to it with him to Bell Labs. But he could not get much traction in using this element there.

In 1947, life at Bell Labs changed that. That December, scientists John Bardeen and Walter Brattain found the basic unit of modern computers, the transistor, an invention that their commanding boss, William Shockley, was also trying to create. Within a transistor, small electrical signals could be amplified. To make it work,

Brattain stabbed a hunk of germanium with two wires, with the germanium resting on a copper base connected to another electrical circuit applying a voltage. What they found is that a weak signal going into one of the wires came out as a much stronger signal from the second wire; the signal came in as a whisper, but came out as a scream. They found that the electricity going across the germanium could also be shut on or off, too, like a faucet, or like a switch.

Bell Labs was looking for a new way to connect and direct telephone calls, replacing the hundreds of female switchboard operators. At the rate that phone calls were increasing in the United States, the Bell Labs brass joked that they were going to need to hire half the "girls" in the country to work at switchboards. In addition, Bell Labs wanted a device that did not wear out like Strowger's contraption. The transistor now could serve as that switch.

On top of a new switch, Bell Labs needed to amplify phone call signals. When Teal was a boy, telephone calls from Texas to New York were not possible, since the telephone signal weakened as it traveled through the copper wires. With the introduction of vacuum tubes, which look like light bulbs with additional parts inside, weak signals were strengthened and long distance calls became possible. With them, Teal could call Mama Teal back home in Texas. However, vacuum tubes were inefficient; they were bulky, they were hot, they hogged lots of power, and they blew out. The transistor amplified the signal, was pea-sized, stayed cool, used less power, and failed less often. The transistor was the electronic age's Holy Grail, and Bell Labs found it. And, at the heart of the transistor was germanium, Teal's favorite element.

An ability to switch electricity in electronics was essential in the world of science. Scientists like Teal and others knew it. Every person within Bell Labs and outside of its beige brick walls wanted to be part of this project. New science, new inventions, and new businesses were now possible. Every scientist clamored to be involved, but this research was in the realm of the physicists and the metallurgists.

These groups were located within the maze of hallways at Bell Labs' Murray Hill campus on different floors in separate buildings from Teal's department. In the scheme of this research lab, the transistor work was a world away from Teal.

Teal tried to plug into this group, offering not only the benefits of germanium, but also those of a flawless form—one single crystal. Crystals that were made up of more than one crystal, or polycrystalline, had surfaces within them, called grain boundaries, looking like a fractured windshield. These boundaries acted like speed bumps; they slowed down the flow of electrical current and caused transistors to behave differently one from another. Electricity would move faster within a single crystal and transistors would operate simlarly. The physicists, however, particularly the head of the group that made the transistor, William Shockley, didn't feel that Teal's single crystals were needed. Shockley, the leader of the Solid State Physics Group that contained both Bardeen and Brattain, tried to control all aspects of the project. Like the Roman official Pontius Pilate, Shockley determined the fate of any new proposal, and he crucified Teal's idea.

On an autumn day in late September 1948, a rare occasion when Teal headed home in time for dinner, he walked from his lab at Bell Labs and down a seemingly infinite corridor to the main entrance to catch the bus to the train station in Summit, New Jersey. The last bus at 5:50 p.m. got him to the station by 6:07 p.m. His home was a short fifteen-minute walk from there. The Summit and New Providence bus line, run by Michael De Corso, was a reliable service for Bell Labs employees. In a cerebral world of research laboratories, reliable infrastructure, like employer-sponsored transportation, allowed brilliant minds to flourish without the worry of how to get home. With only a few hundred riders per day, this transportation business was no moneymaker. But this bus line took home some of the smartest people on the planet on its jade-green upholstered seats. This 1940 motor coach, manufactured by White and shaped like a loaf of bread, got three miles per gallon and shuttled scientists from the inventing world of Bell Labs to the so-called real world. For many, like Teal, their bodies were physically moved but their minds always stayed behind the beige brick facade of Murray Hill.

While waiting for the bus that evening, a fortyish Teal, now with more girth and less hair, stood next to his colleague John Little, a mechanical engineer. While chemists, including Teal, were on a low rung at Bell Labs, engineers were even lower. But John Little was an insider on the transistor project. Little, who was working part-time in Murray Hill, New Jersey, and part-time in New York City, lamented as they walked up the three steps and entered the bus's aquamarine hull about how he needed a small piece of germanium for his transistor project. In that moment, as they both placed their cases in the luggage rack and scooted in their seats past the armrest, Teal's time had arrived. In Teal's cool, calm, and understated style he told Little, "I can make you a rod of germanium." To that he added, "It will be single crystal, too." The germanium would be flawless.

On that short four-mile bus ride, time stopped for these two men. Starting off from Mountain Avenue, this mismatched pair scribbled out plans on small pieces of worn paper from their pockets, and the scenery around them blurred. They cobbled a design of how to make this crystal of germanium for the transistor project.

They decided to form a crystal by pulling it out of a liquid form of the metal, like rock candy on a string in sugary water. Two days later, on October 1, 1948, Teal, ignoring his assigned work, was in New York City on the first floor of 463 West Street in Little's laboratory feverishly putting together the towering equipment. To make it work, they needed to heat germanium to the extreme, so they used heating coils from Little's Lab. They also needed to take all the air out of the environment, so they used a vacuum system to suck it all out. They needed to stop reactions from occurring with germanium, so they attached a gas tank of hydrogen to flow over the crystal. They also needed to gently pull the crystal out slowly from the melt, so they took apart a clock to get its motor.

Their method was to dip a small germanium seed into a palm-sized pool of molten germanium. When the bottom of the cool seed touched the surface of the hot liquid, they stuck to each other like a tongue on a frosty flagpole. As Teal slowly pulled the seed upward, a thin skin of liquid froze on its underside, growing the long crystal layer by layer. Out of this liquid came a long silvery-colored stick, knotted like a tree in parts, and thin as thread in others. Teal mastered the tug-of-war between the liquid and the solid, pulling the germanium crystal from the melt, and germanium, his long-time companion, complied.

This idea of crystal pulling was found accidentally years before during World War I. In 1916, Polish scientist Jan Czochralski was ending his day and finalizing his lab notes with his fountain pen. Absentmindedly, he inserted his pen into a nearby pot of melted tin instead of his inkwell. When Czochralski looked, he found there was a thin thread of metal hanging from the pen's nib. Czochralski discovered a quick, easy, and inexpensive way to make perfect pieces of metal, and went on to become one of Poland's greatest scientists along with Marie Curie and Nicolaus Copernicus. His fame never crossed the Atlantic, but his work did.

Borrowing from Czochralski's method, and working without management's approval, Teal and Little created long slivers of metal the length of a hand. These germanium crystals were flawless on the inside, but ugly on the outside, looking like knotty tree branches. Inside, Teal's germanium crystals were perfect, which was unlike the response from management. When he brought these gems to the attention of physicists, they rejected them. Shockley's words that single crystals of germanium were unnecessary had reverberated down the halls of Bell Labs. Teal knew that Shockley could be "pigheaded." "I thought that was silly," Teal later said, since without single crystals there was "a complete lack of control."

Working around Shockley, Teal looked for ways to be included in the transistor project from the outside in. First, he worked with John Little. Then, he looked beyond his own management, to Jack Morton, head of transistor development and in charge of manufacturing the transistor. Teal emphasized to Morton that if the transistor was to be realized as a commercial product as real switches and signal amplifiers, then the imperfections of nature needed to be removed. In particular, high purity and high perfection allowed devices to be made with more control and also allowed for this new science of germanium, and other elements like it, called semiconductors, to be chartable. The fraternity of physicists was thinking short-term—a one-off proof of principle, and a shot at a Nobel Prize. Teal was thinking long-term—the mass production of switches and amplifiers that were reproducible and reliable. Morton was convinced and funded Teal's efforts, but Teal still had to do his day job in his department.

For most of 1949, Gordon Teal's day had two starts. During the day, he would work on silicon carbide for Bell Labs' new headphones in his lab on the third floor of Building 1; starting at 4:30 p.m., he'd go downstairs to the Metallurgy Lab on the first floor to work on germanium. As the metallurgy department's technicians went home, Teal took out his equipment stored in their closet, connected hefty electrical plugs to power the crystal puller, and attached nitrogen gas, hydrogen gas, water, and vacuum lines to the system. The tower, two feet wide and nearly seven feet high, was taller than Teal's 5'11" frame, making it ornery to maneuver.

All night he'd work out the conditions of making longer crystals, flawless crystals, big crystals. Before the sun came up, he would write down his notes, reverse all the equipment's connections, and cart it back into the closet. The technicians would return a few hours later unaware that their lab was still working when they were asleep.

Teal's wife, Lyda, was mostly supportive of her husband's long hours, but she wasn't happy. Working into the wee hours was romantic when they were in their twenties. Back then, when Teal worked in New York City, she'd walk over meals from their small apartment on Tiemann Place for them both to eat in the lab, and then she would fall asleep on the workbench, while he toiled late into the night. They loved it then. But, now, the Teals had three young boys, who never saw their father. "My family sort of felt that they lost me," he said years later. When Teal's sons did see him, he didn't talk baseball. He talked about germanium and the ability to pull it from a melt.

Gordon Teal pushed past what most scientists were willing to do, like the marathoner he was back at Baylor, by making better and better crystals and giving them to different research groups working on the transistor. Eventually, Teal got his own lab in the same building as the physicists along with an assistant named Ernie Buehler, and filled a room with a mini skyline of crystal pullers. Even Shockley came around to claiming them to be useful. Finally, Teal was able to enter one of the inner circles, working with other scientists, including King Shockley himself. By the end of 1949, everyone in the labs was using single crystals of germanium that Gordon Teal made.

Teal made practical what the physicists had created with a paper clip, gold foil, plastic, and an imperfect gem, by creating a new invention with chemist Morgan Sparks, blending in new ingredients into the liquid germanium. By adding the element gallium in the middle of growing a germanium crystal with other elements in it, the germanium now had two different layers along its length, like a cake, layers that acted differently electrically, making what was called a pn junction. Now, the transistor existed in a more useful form than the science project it once was. Teal had come closer to his glory. But it was still hard to be a star within a constellation of bigger ones. At the end of December 1952, Teal did something about it, saying goodbye to his fellow Bell Labs colleagues and to the East Coast, and ventured off to the wide, open spaces of his home state of Texas. He started a new job at a small company recently renamed Texas Instruments—a day, no doubt, for which his mother had prayed.

Years later, on May 10, 1954, Gordon Teal, a scientist from a little-known company, was scheduled to speak at an electronics conference, hosted by the Institute of Radio Engineers in Dayton, Ohio. During the morning talks, attendees from big companies of RCA, Western Electric, General Electric, and Raytheon heard speakers repeat over and over how impossible it was to make a transistor made of silicon. Silicon was a more robust cousin of germanium, but making a gem of it was challenging. With these sentiments stirring about, Teal waited for his time to present, with his hands in his pockets, as these engineers wrung theirs.

When Teal eventually spoke, somewhere in the middle of his presentation his colleague, Willis Addox, wheeled out a record player. The audience perked up and turned their heads, listening to the unusual pairing of clarinet and harpsichord that rang out from the 45 rpm record of Artie Shaw's "Summit Ridge Road." Plugged into the side of the record player was a paddle with circuits fixed on the flat part, looking like a bionic spatula. Teal announced to the crowd that this circuit contained a germanium crystal and then dipped the paddle with the circuit into hot oil. The clarinet and harpsichord of the record were overcome with static. The scientists weren't alarmed. They knew of germanium's dark secret—it was unstable when it got hot.

So Teal started the demonstration again. This time he had another circuit attached to a paddle. He played the music and dipped this second paddle in the hot oil. When he did that, the clarinet and harpsichord sounded out with no static, over which Teal announced that a silicon transistor was playing music for them at that very moment.

One person sitting in the middle of the auditorium jumped up and asked Teal, "Do you have silicon transistors in production?" to which Teal flatly replied, "I just happen to have some here in my coat pocket." As he uttered those words, he put his hand into his pocket and pulled out a small metal gizmo that looked like a three-legged robot from a science fiction movie. The future had arrived.

One person dashed to a public phone and yelled into the line, "They have silicon transistors in Texas!" A metallurgical miracle had materialized and Teal received his overdue uproar. But more important, the computer, like the Tin Man from *The Wizard of Oz*, got its brain. With silicon transistors, the computer got the part that allowed it to compute and to think, by acquiring a basic building block that when working together would do more, like the switches on Coy's switchboard, but on a vastly grander scale. This building block of a transistor—a silicon-based switch—when combined with others, would not only make computers smarter, surpassing humans, but also switch how humans think.

Shaping the Brain

As Coy, Strowger, Teal, and many other scientists used the technologies before them to make better switches, these switches became the heart of telephone systems and later computers. But the creation of these switches let to the re-creation of our brains as well. Computers are affecting how humans think. Early computers took on simple tasks with the intention of augmenting human cognition, and with great scientific and engineering advances, eventually these computers matured to allow creation of the world wide web of the internet. But the development of transistors has been a major factor in computers becoming ubiquitous, allowing the internet to reach everywhere and touch everything. The world before computers is different from the world after them, and scholars are examining life that includes the transistor, the computer, and the internet, and asking questions. What is certain among these scholars is that these technologies are molding our brains right now.

While scholars all agree that the reach of the internet has extended to our brains, where there is a debate is if the internet is making us smarter or dumber. The answers to that question are "it is hard to know" and "it depends on who you ask." Whenever scientists want to do an experiment, they need one group to undergo a change and another group to stay the same for comparison. The second group is called the control, and it serves as a kind of pace car to see what the experiment is actually doing. When it comes to testing the impact of the internet, it is incredibly difficult to find anyone who hasn't been on it. It is hard to find a control. Those who can be controls might have other factors that make them a bad comparison, such as speaking a different language, or suffering from poverty, or living in a different culture, like the Amish. Be that as it may, this dilemma doesn't stop scientists, scholars, and citizens from making assertions or speaking from intuition on the impact of the internet.

In one camp are the optimists, who say that the internet is making us smarter. In just a few clicks of a mouse the data that we desire shoots on a beam of light across an optical fiber straight to our computer screens. In the same amount of time it takes a hummingbird to flap its wings, we are able to answer questions like: Where is Timbuktu? What is the capital of Utah? How many feet are in a mile? Just a few decades earlier, answering these questions would have taken more time than a pizza delivery. In the past, we would have to get a map, or open an encyclopedia at the library, or pull out a conversion table and a calculator. "The Internet is just a terrific way for suddenly getting exposed to an entire planet's worth of ideas," said neuroscientist David Eagleman. "I think it's definitely not making us dumber," he said. "I think it's going to make us much, much smarter."

Others, however, are not so optimistic about the impact of the internet. Since 2008, essays like Nicholas Carr's piece in the *Atlantic Monthly*, "Is Google Making Us Stupid?," highlighted the internet's red flags, which were noticeable ten years after Google's founding and eighteen years after the internet's birth. Carr discussed in his follow-up book—*The Shallows,* on how the internet is changing our brains—that the web, with its smorgasbord of disjointed facts, its buffet of media types (of text, pictures, video, and audio), and its dash of links, serves up information in a raw and unprocessed form that our brains must digest and incorporate. This strains our brains. From centuries of knowledge acquired from books, our gray matter is accustomed to linear thinking, as one idea flows into the next and into the next. With the web, however, ideas don't flow; they yank, tug, jerk, and jolt. Additionally, with the inundation of information, we have developed new reading habits to accommodate the deluge. When reading webpages, we get what we need by skimming the text, by hunting for key words, and by scanning superficially. With these new habits, neuroscience tells us that our brains become more proficient at these skills. Some scientists and scholars believe that the learning habits we develop from our use of the internet weaken our ability to think deeply.

What scientists have learned is that our memory has parts: the short-term memory and the long-term memory, where the former holds information for seconds, and the latter for years. But there is also the working memory, which connects them, acting as scratch paper where ideas retrieved from the long-term memory are worked out. When figuring out the tip for a meal, or remembering the next step in a recipe, or rotating an object in one's mind, all that is happening in the working memory.

The working memory has a limit to how much it can hold. The telephone company learned about this limit in the early twentieth century, as it was on a quest to make better switches. As phones became popular in the 1920s, telephone numbers were seven digits in order to mathematically produce unique phone numbers for people in big cities. Seven would be a fortunate choice.

Telephone numbers were not all numbers at first. They were part letters and part numbers; for example, a New York number might be "PEN(nsylvania) 5000." But, by the 1960s and 1970s, all phone numbers were seven digits. However, a problem arose. People were misdialing them because they were remembering phone numbers incorrectly. Bell Labs sponsored a study on the working memory's capacity for long strings of numbers like 15553141593. The study discovered two important things: they found that if the number was chunked like 1-555-314-1593, phone numbers could be remembered correctly (which is why US phone numbers look the way they do). They also found that the working memory is much like an express checkout line, with a limit of items it can handle. For the working memory, the magic number is seven items (more or less).

Our consumption of information is constrained by the working memory. As Carr explains in *The Shallows,* the working memory carries cupfuls of information scooped to and from the pool of the long-term memory. The internet, however, is Niagara Falls. Additionally, the random splashes—of a video here, a quick fact there, a Facebook post here, a tweet there—get jumbled up into our working memory, which is transferred to our long-term memory, so we end up scatterbrained.

We have entered a time where we are overloaded with attention-grabbing headlines and clickbait, and don't have the space to think deeply as we flitter from one story to the next. Our knowledge resides on the surface. If we were to read a book, we would be fully submerged in the details and nuance of another world and swim in the deep end. The internet, however, is a world wide wading pool.

We slosh in the superficial, because we have reached a critical point for what our brains can hold, which is forcing society to have a new relationship with information. This impasse has been depicted in an artful way by Hollywood, showing us what the internet has become for us. In the 2001 movie *Memento,* a man named Leonard Shelby desires to find his wife's murderer. But there's a catch. Leonard has a condition causing him to be unable to store new memories (an illness called anterograde amnesia). Cleverly, with an assortment of items, Leonard rustles up a rudimentary way to remember. In his pocket are Polaroid pictures (of his hotel and people he "knows"). On his chest and arms are tattoos reminding him of the facts of the case ("the murderer is John or Jim G"). On his wall is butcher paper, where the Polaroids are stored and labeled. Together, these serve as his memory, residing outside of his body, since his biological memory (his brain) is impaired. His memory aids would be defined by philosophers as an extended mind. With them, "roles the brain used to play could get taken over by tools in the world," said NYU philosophy professor David Chalmers, who cowrote the paper "The Extended Mind" in 1998. Leonard's mind is not within his skull, but beyond it. This notion of the extended mind is not just a heady academic concept from a decades-old philosophical paper. This concept of the extended mind became a prediction. The internet has indeed become an extended mind for all of us.

Extended minds need not require a visit to the tattoo parlor or the craft store. In hindsight, we understand now that, by definition, humans have had extended minds on a small-scale for generations. Carvings on ancient walls, clay tablets, scrolls, and books are all elements of an extended mind, but so are grocery lists and Post-It notes, calendars and checklists. To be considered an extended mind, a few criteria must be met. "We get it, we trust it, we use it," said Chalmers. The items must be accessible, trustworthy, and available, and with the internet in our pockets, the web now fits that definition.

Researchers have found that the internet has changed our behavior so that memories that would have been precious to us in our predigital days, like our mother's phone number, aren't precious any longer. Scientists have proven that we no longer remember things; we now remember where they are located. Instead of remembering a phone number, we know to get a cellphone and give a command to retrieve it. When it comes to information, our brains prioritize "where" over "what." We don't need to remember, when apps will do it for us. As such, our brains have been rewired by the internet. We've turned into Google Brains.

In ancient times, cultures passed down their histories with memorized oral traditions. In early schools, students recited poems or the Gettysburg Address or a list of the state capitals by heart. Not long ago, we used to know phone numbers, which serendipitously suited our working memories. Those traditions are gone. For some, not remembering phone numbers is a sign of our advancing times. "It's not an important loss," said neuroscientist Eagleman.

There are some benefits to this human–computer symbiosis. The abacus, Charles Babbage's mechanical computer, Ada Lovelace's software, the ENAIC, and the integrated circuit made calculations easier. The ability to have something do computations was a good thing, since the brain is not a very good calculator. The web has become an extension of the mind on a scale larger than anything humankind has ever experienced, though, and with it we adopt the posture that we do not need to know. The ease of looking things up assaults our ability to understand and to experience. There is a difference between experientially-knowing and YouTube-knowing. Knowledge cannot be found in a Google search. Wisdom cannot be found in an algorithm. Understanding cannot be downloaded.

Knowledge, wisdom, and understanding are not the only aspects of humanity touched by the internet. Creativity is, too. The brain and how it creates is another mystery of neuroscience, but scientists have identified parts of the brain that grow with different creative endeavors. Musicians have an increase in one part of their brain; visual artists in another; and writers another. There isn't much certainty in how the brain creates, but there is certainty that it is influenced by the internet. There are two opposing sides on how the internet affects creativity, however. If creativity is defined as the blending of ideas, the breaking of ideas, and the bending of ideas, with that, the internet can help one become more creative. That's what neuroscientist David Eaglemen believes. "The more of the world you absorb, the more creative you can be because you just have more raw materials to then break and blend," he said. Creativity can also have several steps to it, of preparation, innovation, and production. The web is a great tool for the first step. "The Internet can provide the investigator information more rapidly," said Professor Kenneth Heilman, Professor Emeritus of the Department of Neurology at the University of Florida.

But there is a downside. Creativity is not just the warehousing of ideas, but a process of giving the brain time to simmer on these ideas. Creativity requires preparation, but it also needs incubation. "When a person is alone and relaxed, they often come up with very creative ideas," wrote Heilman. One classic example is Sir Isaac Newton sitting under the apple tree. "Perhaps if Newton was reading his email at this time," wrote Heilman, "he would not have developed his creative ideas." Newton might not have even seen the apple fall if his attention was fully absorbed by his smartphone.

According to Eagleman, there are two parts to being creative: "absorbing the whole world" and "having the time to digest and put things together in a new way." Attaining the second part in our technological age isn't easy. Our time with our technologies is counter to creativity. Even cyber-optimists like Eagleman agree. "Obviously," he said, "there are a thousand ways to waste time on the internet." Our time on the web and our penchant for multitasking floods our brains. Additionally, when our working memories are at their maximum capacity, we are more susceptible to being distracted. Our distractions become more distracting and these distractions, along with the addictive nature of the web, prevent us from maximizing the web's potential. Plus, we have old brains living in a modern world. Our hunting and gathering minds exist in an age where there is nothing to physically hunt or gather, so our brains get trapped in the cycle of hunting and gathering "follows" and "likes" on social media. The internet might be a tool to enhance deep thinking, but the way we use it—with its tremendous number of distractions—doesn't make us deep thinkers.

We know that we are at a crossroads, and even the makers of this technology know that while something is being gained, something else is being lost. In many private schools in Silicon Valley, where money is no object, visitors will be troubled to find something missing. There are no computers! Some parents in Silicon Valley prevent their children from using the technologies that they helped to bring into the world. Even Steve Jobs, the father of Apple, was a "low-tech dad." Some cyber-optimists believe they know why there is a resistance. "I think it's just a fear of new things," said Eagleman. For centuries, there has always been a backlash against the new. In ancient Greece, scholars complained that writing made students less able to remember oral traditions, and thereby less intelligent. Computers might be the high-tech version of that concern. "It's a big question of finding that middle ground," said NYU philosophy professor Chalmers.

We certainly gain something with our technologies. Research shows that during the twentieth century, IQ test scores rose every year. We are smarter than our grandparents and our parents. We know more. We can do more, too. However, this ability is not new. "We are doing in minutes what our grandfathers could not have done in days," said Thomas Edison in the nineteenth century. Today, though, we can hear from experts in their TED talks and be up-to-date on the latest in all fields within eighteen minutes. With the internet we have become the "neighborhood" that Samuel Morse predicted.

But something escapes us, too. "The worry for me is that maybe in some cases you might begin to lose understanding," said philosopher Chalmers. "I certainly wouldn't want to get into a situation," he said, "where everything that a child does goes through the computer and a computer is the primary focus."

Like Phineas Gage, we are what our brain experiences. When we continuously use the parts of our brains that require only superficial understanding, we, too, become shallow. When we don't train our brains with deep thoughts, we will eventually not be able to understand, we will not be able to create, we will not be able to think.

The internet, our devices, and our computers are bringing up the question of the value of being human, because what is important to an algorithm doesn't converge with what is important to us. The web knows how fast it can search, how many items come up from the search, and what is the top item of the search. What is vital to humanity does not relate to them. Algorithms care nothing about the quality of our sleep, holidays, language, empathy, bias, scientific breakthroughs, fireflies, the night sky, privacy, or even how humans think. As such, we cannot ask technology to solve these things, for technology cannot see the value of these intangibles. The ingredients that make life worth living, such as music, movies, a great meal, friendship, laughter, justice, peace, stories, festivals, a chance meeting, flowers, traveling, a handwritten letter, love, truth, sports, fashion, hugs, sunrises, sunsets, vacations, fiction, caffeine, and books, mean nothing to a computer. These are all human undertakings and so they require human action to keep them—and protect them.

The computer's processor was originally based on the human brain, but now we are growing to resemble our computers. Aspects of mortals will not fully map onto machines, though. Our gray matter is more complex than a set of switches that swiftly make "yes" and "no" decisions using sophisticated software. Our brains hold the mystery of our genius, our creativity, and our imagination. We are flawed and inefficient, but we are also flexible and intrepid. We do things that seem illogical, but we also innovate. We can create chaos, but we can also create beauty.

The computer's rise forces us to think hard and deeply about what really makes us human. There is a fork in the road and humanity must decide on its guiding principal—the making of better machines or becoming a better species. This moment asks us to consider our trajectory. This moment is also calling on us to be brave. If we don't like the direction we are headed, we must be courageous enough to change bearings and take a different tack.

We must be bold enough to make a switch.

Epilogue

The writing of *The Alchemy of Us* was bookended by two quotes attributed to Nobel Laureate Toni Morrison. The first of these quotes was one I was familiar with from the beginning of this project, for it served as the book's catalyst: "If there is a book that you want to read, but it hasn't been written yet," said Morrison, "you must be the one to write it." From my experience as a black woman scientist, I often found that my reflection in textbooks was hidden, missing, overshadowed, or cast in a poor light. When the opportunity came for me to write about science and technology, I heeded Morrison's words.

Admittedly, when I began writing this book, I initially subscribed to the established ways of thinking about science and technology, and was bound up with the retelling of old stories of white men and their inventions. But the writing of this book led to my own alchemy. A surprising fire shut up in my bones halted me from writing stories where my own reflection was missing. In addition to this visceral response, I also grew to understand the power of reflection and that every reader needs to see themselves in stories. As such, I attempted to create mirrors within the text. The inventors highlighted in this book had talents, but also shortcomings, which are ingredients we all have. So in these pages, I tried to uncover and display their complexity and their humanness, so that readers— whether in the sciences or not, whether of the same demographic as the inventors or not—can connect with these characters on some level and see something familiar. I would have loved to have such a book while taking my engineering courses so long ago. One set of books in my backpack would have filled my head, but this book would have fed my soul.

Books about technology do not often employ such a strategy of making inventors human. Many authors want to glorify genius, but by doing so inadvertently make innovation seem unachievable. On the other end of the spectrum, scholars want to share as much as they can. Perhaps that academic approach is best in some cases, as in the esteemed work of Lewis Mumford, Jacques Ellul, and Thomas Kuhn. But I made the conscious decision to sacrifice the amount of technical and scholarly content, which might be understood by the few, and opted instead to tell human stories, which can be readily absorbed by the many. Although I wasn't sure about this approach in the beginning, Professor Morrison convinced me why this strategy was a sound one in a second quote.

While puttering around after completing my first draft, I stumbled onto a keynote address given by Morrison that transformed my uncertainty into certitude. In 1991, the soon-to-be Pulitzer Prize winner discussed the importance of including different vantages, experiences, and cultures in the academy. She challenged professors "to re-read traditional texts of a discipline with a fresh point of view," and she explained that such efforts would uncover, "more rather than less power, more rather than less beauty, more rather than less intellectual vigor—and subtlety." Not doing so, she warned, would lead us to the "very Dark Ages." Morrison's speech told me that when it came to these well-known stories, my vantage and my approach weren't inconsequential, but imperative.

Her words resonated with me on many levels, because while writing this book I encountered an alarming story of how the best of intentions to encapsulate culture almost went catastrophically wrong. I found that several decades prior, in 1977, Carl Sagan and friends had a rare opportunity to launch a record into space during NASA's Voyager mission. This ad hoc committee worked hard to find the songs of Earth to be included in the short ninety minutes of play time. Sagan, a forty-three-year-old white man and a classical-music lover, initially chose selections that were largely of European origin. After much effort, the younger members of his team were able to pepper in music from other cultures. It wasn't until they consulted Alan Lomax, a longtime collector of the world's songs, however, that their playlist represented the entire planet. Lomax resonated with me, because if he hadn't been involved, the Golden Record—an interstellar time capsule of Earth—would have portrayed only a fraction of the planet.

While there isn't a shortage of books on science and technology, many writers view their work from their own lens, just like the beloved Sagan. I have attempted in these pages to apply a Lomaxian approach to thinking about this subject, an approach that Professor Morrison assured was proper. Discussions about technology must be inclusive, because technology is neither just for the few who are learned, nor is it just for men who are of European descent. Everyone makes something, from a sandwich to a solar cell, so examinations of science and technology must reflect this. Every person can create something new, whether it is splicing beats with two turntables and a microphone or splicing genes with two test tubes and CRISPR. As such, stories about science and technology must reflect that innovation is universal.

When books about technology reflect readers, those readers come away with more than just stories, but a sense that they can create,

Acknowledgments

Thinking back on the process of writing this book, I am filled with gratitude for the opportunity to create it and for the people who cheered for me along the way. My mother, Angela Pitaro, gets my biggest thanks, for she believed in me and in this project even when I didn't. Thanks also go to my brothers, Davyd and Marc; my niece and nephew, Lena and Alex; and my sister-in-law, Cassandra, for their steadfast support and love. In addition to my family, my friends have been amazingly supportive, too. A special thanks goes to my dear friend, Robin Shamburg, who served as this book's midwife and top cheerleader. I am also grateful for Sarah Marxer, whose friendship feeds my soul; and, for Kathy Yep, my faithful running buddy. In addition, Gina Barnett, Wendy Sealey, Ines Gonzales, Katherine Vorvolakos, Leslie Kenna, Emilie Lorditch, Gina La Cerva, and Erin Lavik made this long journey less onerous. Mildred Mewborn, Lamont White, Ron Knox, Philip Fiondella, Nancy Santore and Victorio Sweat made the days more pleasant too. And my former students, particularly, Katie McKinstry, Guy Marcus, Jeremy Poindexter, and Xu Huang, inspired me to forge new ground.

My love for science was nurtured long ago by a public television show, but also by great science teachers. They include: Kathleen

Donohue, Jean-Marie Howard, and Dr. Edelgard Morse. If Dr. Morse didn't have her Chemistry 21T course at Brown, my dream of becoming a scientist would have been deferred. Thanks, Dr. Morse! In addition to science, my teacher Ms. Aspaia Verpuit awakened my love for history.

I am grateful for the mentoring and support from Shirley Malcom, Anne Fausto-Sterling, Samuel Allen, Clayton Bates, James Mitchell, Lisa Marcus, Usha Kanithi, David Johnson, Jr., and Paul Fleury. I am also thankful for the Jodi Solomon Speakers Bureau for helping me spread my message that science is fun—and for everyone. Lastly, it has been a pleasure and privilege to work with the MIT Press. My deepest appreciation goes to my editor, Bob Prior, for being so supportive (and patient) and to its director, Amy Brand, who encouraged me from the outset.

There is an old African proverb that says that "it takes a village to raise a child." I've learned that that expression holds true for producing a book as well. I have had support, assistance, and encouragement from many wonderful organizations and kind souls. In acquiring materials for this book from across the pond, Angela Pitaro (my mother) turned her vacations to England into research trips. I am grateful for the pictures and materials she acquired, but also for this new bond between us. For materials acquired in the United States, my thanks go to Jo Chapman for securing materials from Californian archives and to Kasja Spanou and Alba Morriss for tracking down hard-to-find journal articles. My thanks go to Dulsie Rebecca Feurtado for her transcriptions, to Mark Saba for the artwork, to Bev S. Weiler for the copyediting, and to Michael Sims for editing my final draft.

One of the best gifts a person can give a writer is a critical ear. Hilary Brueck, Carey Reed, and Nick Smith have been exceedingly generous toward me and I am grateful for their unwavering support and encouragement. A special thanks goes to Prof. Sam Freedman for allowing me to audit his class in 2014, as well as to Kelly McMasters, who helped me fashion this book at its genesis, and later helped me polish the final draft. My appreciation also goes to Prof. Robert Gordon of Yale, who provided thoughtful comments on the scientific content, as well as to Marie Brown for ushering my book through the initial negotiations.

There were numerous archives that helped make this book possible. They are all listed elsewhere in the book, and I am grateful for all of them. However, there are a few individuals who went beyond the call of duty. They include: Sheldon Hochheiser (AT&T Archives), Melissa Wasson (AT&T Archives), William Caughlin (AT&T Archives), Ed Eckert (Nokia Archives), Rebecca Naldony (Nokia Archives), James Amemasor (The New Jersey Historical Society), Gordon Bond (Garden State Legacy), Trina Brown (Smithsonian Libraries), Charlotte Chapel (Friends of Plume House), Jamie Martin (IBM Archives), Kenneth McNelis (The Museum of Bus Transportation), Paula Norton (The Derby Historical Society), Sara Paramigiani (The Fleetwood Museum), Kay Peterson (The Smithsonian Institute), David Rose (The March of Dimes Foundation), Edward Sax (The Vintage Radio Museum), the late Charles Seccombe (Ansonia, CT), Daryl Smith (Yale Glassblowing Lab), Frances Skelton (The New Haven Museum), Ed Surato (The New Haven Museum), and Hal Wallace (The Smithsonian Institute). There were scores of interviews taken for this book, but I am particularly grateful for the generosity of John Casani, Frank Drake, Timothy Ferris, Caroline Hunter, Nancy Marrison, David Rooney, and Dr. Donald Teal.

Special thanks goes to my local library—the New Haven Free Public Library—particularly the incredibly supportive staff at the Mitchell Branch. A special shout out goes to Sharon Lovett-Graff for her skill in obtaining impossible books and references. Thanks to Seth Godfrey too, for giving me access to the library's closed stacks. I am also grateful for the space to write within the walls of Buley Library.

Acquiring materials for this book was certainly a blessing, but so too was the financial support I received. I am deeply appreciative for the generosity of the following institutions: Thanks to The Texas Collection of Baylor University, and its director John Wilson, for the travel fellowship to visit the Gordon Teal Papers. Thanks also to the Djerrasi Resident Artists Program for the extraordinarily productive and enriching month of writing in July 2017. Last, but certainly not least, I would like to thank the erudite and supportive Doron Weber and his staff at the Alfred P. Sloan Foundation for the book grant, which allowed me to create the best possible prose I could produce, and enabled me to secure illustrations and photographs. They say "a picture is worth a thousand words." For each, I offer one thousand thanks.

To be honest, there are just too many people to list who have been good to me during the time of this book project. So let me just close with a broad but heartfelt, "Thank you!"

Notes

Sources Collections and Interviews

The writing of this book benefited from the following libraries, archives, and collections:

Alexander Fleming Laboratory Museum, AT&T Archives and History Center, Baker Library of Harvard Business School, The Bancroft Library, The British Library, Cambridge University Library, Chicago History Museum Archives, Columbia University Archives, Computer Museum Archives, Connecticut State Library, Corning Inc. Archives, DeGoyler Library of Southern Methodist University, Derby Historical Society, The Fleetwood Museum of Art and Photographic, Friends of Plume House, George Eastman Museum, George Washington University Archives, The Henry Ford, History Center in Tompkins Country, IBM Archives, IEEE Historical, Ironwood Area Historical Society, Ironwood Carnegie Library, Kansas City Public Library, Kansas Historical Society, Kingston Museum and Heritage Service, LaPorte Historical Society, Library of Congress, Michigan State University, Morristown & Morris Township Library, The Museum of Bus Transportation, Museum of Science and Innovation Archives, Napa County Historical Society, National Archives at

Kansas City, New Haven Free Public Library, New Haven Museum, New Jersey Historical Society, New York Historical Society, New York Public Library Archives and Rare Books Division, Newberry Library, Niels Bohr Library & Archives, NOAA, Nokia Archives, Penfield Historical Society, Rakow Research Library at the Corning Museum of Glass, Royal Society Library, San Francisco Public Library, San Jose Public Library, Schomburg Center for Research in Black Culture-Manuscripts, Schott Archives, Science History Institute, Smithsonian Institute, Southern Connecticut State University Library, St. Petersburg Florida Library, Stanford University Archives, The Texas Collections at Baylor University, Trinity College Library, Ulysses Historical Society, Ulysses Historian Office, Union College Archives, Vintage Radio and Communication Museum, Waco-McLennan County Library, The Historic New Orleans Collection's William Research Center, Worcester Polytechnic Institute Archives, Xavier University of Louisiana Archives, and Yale University Archives.

The writing of this book benefited from interviews from the following individuals:

Gretchen Bakke, John Ballato, Naomi Baron, Roger Beatty Fernando Benadon, Paul Bogard, Marvin Bolt, Gordon Bond, Kevin Brown, John Casani, Robert Casetti, David Chalmers, Oliver Chanarin, Charlotte Cole, Jane Cook, Leo Depuydt, Frank Drake, Nancy Jo Drum, David Eagleman, Joanna Eckles, A. Roger Ekirch, Fabio Falchi, Isobel Falconer, Timothy Ferris, Mariana Figueiro, Ariel Firebaugh, Robert Friedel, Peter Galison, Jon Gertner, Robert Gordon, Kenneth Heilman, George Helmke, Albert Hoagland, David Hochfelder, Sheldon Hochheiser, Caroline Hunter, William Jensen, James Jones, Kathy Kanauer, Art Kaplan, Daniel Kelm, William LaCourse, Ed Lax, Robert Levine, Sarah Lewis, John Littleton, James Lloyd, Travis Longcore, Bertram Lyons, Nancy Marrison, Avalon Owens, Mark Rea, Susie Richter, David Rooney, Wolfgang Schivelbush, Daryll Smith, David Smith, Joel Snyder,

Carlene Stevens, Richard Stevens, Donald Teal, Leslie Tomory, Susan Troilier-Mckinstry, Geoff Tweedale, Hal Wallace, Thomas Wehr, Wayne Wesolowski, Matthew Wolf-Meyer, Randall Youngman, and Evitar Zerubavel.

Source Notes

ATT	AT&T Archives and History Center
AS-PEN	Almon Strowger Vertical File, Penfield Historical Society
AV-SI	Vail Telegraph Collection (Record Unit 7055), Smithsonian Institute Archives
BAY	Gordon Kidd Teal Papers, Accession #3820, The Texas Collection, Baylor University
CMOG	Rakow Research Library, Corning Museum of Glass
COR	Corning Inc. Archives
CSL	Connecticut State Libraries
DHS	William Wallace Vertical File, Derby Historical Society
HH-LC	Herman Hollerith Papers (MSS49510), Manuscript Division, Library of Congress
HG-NJHS-VF	Hannibal Goodwin Vertical File, The New Jersey Historical Society
HG-NJHS-PELL	Papers of Charles H. Pell (MG1041), The New Jersey Historical Society
JJH-WPI	Jacob Hagopian Papers (MS13), Worcester Polytechnic Institute Archives
JJT-TRI	Papers of J. J. Thomson, Trinity College Library, Cambridge
MUY-SUL	Walter R. Miles Research concerning Eadweard Muybridge (M0736), Department of Special Collections, Stanford University Libraries

NHFPL	New Haven Free Public Library, Connecticut
NOHC	War of 1812 Newspaper Collection (Mss 499), Williams Research Center, The Historic New Orleans Collection
POL-HBS	Polaroid Corporation Administrative Records, Baker Library Historical Collections, Harvard Business School
PRWM-SCH	Southern Africa Collective Collection (Papers of the Polaroid Revolutionary Workers Movement, PRWM), Schomberg Center for Research in Black Culture, The New York Public Library
RJ-CHM	Rey Johnson Papers (Lot X3312.2006), Computer History Museum
SAG-LC	The Seth MacFarlane Collection of the Carl Sagan and Ann Druyan Archive, Library of Congress
SFBM-ONE	Edward Lind Morse, ed. *Samuel F. B. Morse: His Letters and Journals*, Vol. 1: Houghton Mifflin Co., 1914
SFBM-TWO	Edward Lind Morse, ed. *Samuel F. B. Morse: His Letters and Journals*, Vol. 2: Houghton Mifflin Co., 1914
SFBM-YUL	Morse Family Papers (MS 359), Manuscripts and Archives, Yale University Library
TAE	*The Papers of Thomas A. Edison*, Volume 3 published by The Johns Hopkins University Press, 1989
TAE-RU	The Thomas Edison Papers, Rutgers, The State University of New Jersey (edison.rutgers.edu)
WUTC-SI	Western Union Telegraph Company Records, Archives Center, Smithsonian Institution

Chapter 1: Interact

1 **just like every Monday** David Rooney, *Ruth Belville: The Greenwich Time Lady* (London: National Maritime Museum, 2008), 91.

1 **"four seconds fast"** Donald De Carle, *British Time* (London: C. Lockwood, 1947), 108.

2 **license and livelihood** Rooney, *Ruth Belville*, 64.

Reaching its gate Rooney, telephone interview by author, March 4, 2016.

3 **"vice of speed"** Robert James Forbes, *The Conquest of Nature: Technology and Its Consequences* (New American Library, 1969), 118.

3 **after a designated appointment time** Robert Levine, telephone interview by author, May 2, 2016.

3 **two separate intervals** A. Roger Ekirch, "The Modernization of Western Sleep: Or, Does Insomnia Have a History?" *Past & Present* 226, no. 1 (2015): 156.

3 **sleep for three and a half** Ekirch, telephone interview by author, April 22, 2016.

4 **clean, or gossip** Ekirch, "Modernization of Western Sleep," 152.

4 **Ancient texts** Ekirch.

4 **number of classics** Ekirch, 158.

4 **sleep hundreds of times, too.** Ekirch, telephone interview.

4 **invention of artificial lights** Ekirch.

5 **clock gave it its pulse** Edward P. Thompson, "Time, Work-Discipline, and Industrial Capitalism," *Past & Present*, no. 38 (1967): 82.

5 **creation of new words** Etymologies from the website: www.etymonline.com.

6 **nearly two hundred customers** Rooney, telephone interview.

6 **Maria Elizabeth** Rooney, *Ruth Belville*, 35.

6 **Arnold No. 485** John L. Hunt, "The Handlers of Time: The Belville Family and the Royal Observatory, 1811–1939," *Astronomy & Geophysics* 40, no. 1: 1.26.

6 **a son of George III** "Taking the Time Round," *Yorkshire Post and Leeds Mercury* (Leeds, UK), December 13, 1943, 2.

6 **"warming pan"** Hunt, "Handlers of Time," 1.27.

7 **handy clockmaker** Kenneth Charles Barraclough, *Benjamin Huntsman, 1704–1776* (Sheffield, UK: Sheffield City Libraries, 1976), 2.

7 **tools to rotisseries** Samuel Smiles, *Industrial Biography: Iron Workers and Tool Makers* (Boston: Ticknor and Fields, 1864), 136.

8 **bits of charcoal** Kenneth Charles Barraclough, "Swedish Iron and Sheffield Steel," *Transactions of the Newcomen Society* 61, no. 1 (1989): 79–80.

9 **buried outside his works in Doncaster** Smiles, *Industrial Biography*, 137.

9 **pots from Holland** Alan Birch, *The Economic History of the British Iron and Steel Industry, 1784–1879: Essays in Industrial and Economic History with Special Reference to the Development of Technology* (London: Cass, 1967), 301.

9 **eight to ten hours** John Percy, *Metallurgy: The Art of Extracting Metals from Their Ores, and Adapting Them to Various Purposes of Manufacture* (London: John Murray, 1864), 835.

10 **royal jeweler** Rooney, *Ruth Belville*, 99.

10 **the status symbol of GMT** Stephen Battersby, "The Lady Who Sold Time," *New Scientist* 25(2006): 52–53.

10 **hale and hearty** Rooney, *Ruth Belville*, 100.

11 **between noon and 2:00 p.m.** Ed Wallace, "They're Men Who Know What Time It Is," *New York World-Telegram*, December 23, 1947, 17. *ATT*.

11 **Inverary, Ontario** W. R. Topham, "Warren A. Marrison-Pioneer of the Quartz Revolution," *Bulletin of the National Association of Watch and Clock Collectors, Inc.*, no. 31 (1989): 126–134.

12 **he explained science to her** Nancy Marrison, telephone interview by author. March 24, 2016.

13 **"measuring stick"** Warren A. Marrison, "Some Facts About Frequency Measurements," *Bell Labs Record* 6, no. 6, 386.

15 **ME7–1212** *The World's Most Accurate Public Clock* (pamphlet) (New York: American Telephone and Telegraph, 1941), 1. *ATT*.

15 **"are the only species"** Matthew Wolf-Meyer, Skype interview by author, May 2, 2016.

15 **plunged seven males** Thomas A. Wehr, "In Short Photoperiods, Human Sleep Is Biphasic," *Journal of sleep research* 1, no. 2 (1992), 103–107.

15 **"older pattern of sleep"** Ekirch, telephone interview.

15 **sleep disorders or sleep deprivation** Bruce M. Altevogt and Harvey R. Colten, ed. *Sleep Disorders and Sleep Deprivation: An Unmet Public Health Problem* (Washington, DC: National Academies Press, 2006), 1.

16 **sleep-aid prescriptions** Yinong Chong, Cheryl D Fryer, and Qiuping Gu, "Prescription Sleep Aid Use among Adults: United States, 2005–2010," *NCHS Data Brief*, no. 127 (2013): 1–8.

16 **"in all of history"** Ekirch, telephone interview.

16 **sleep deprivation with rats** Allan Rechtschaffen et al., "Physiological Correlates of Prolonged Sleep Deprivation in Rats," *Science* 221, no. 4606 (1983): 182–184.

16 **loss in brain function** Michael A. Grandner et al., "Problems Associated with Short Sleep: Bridging the Gap between Laboratory and Epidemiological Studies," *Sleep Medicine Reviews* 14, no. 4 (2010): 239–247.

17 **methods to synchronize clocks** Peter Galison, *Einstein's Clocks and Poincare's Maps: Empires of Time* (W. W. Norton, 2004), 248.

17 **embodiments were workable** Galison, Skype interview by author, May 2, 2016.

18 **about four minutes later** "Time's Backward Flight," *New York Times*, November 18, 1883, 3.

18 **Illinois, twenty-seven** Carlton Jonathan Corliss, *The Day of Two Noons* (Washington, DC: Association of American Railroads, 1942), 3.

18 **corralled into one system** "Standard Time," *Harper's Weekly* 27, no. 1410 (1883): 843.

20 **fans in the bleachers** Galison, *Einstein's Clocks*, 271.

21 **"Dive and the Prison"** Robert Goffin, *Horn of Plenty: The Story of Louis Armstrong* (Boston: Da Capo Press, 1947), 17.

21 **few hundred milliseconds** Fernando Benadon, "Time Warps in Early Jazz," *Music Theory Spectrum* 31, no. 1 (2009): 3; email to author, April 3, 2016.

21 **"in the spur of the moment"** Louis Armstrong, interviewed by Ralph Gleason, *Jazz Casual*, January 23, 1963, video, 12:52, https://youtu.be/Dc3Vs3q6tiU

21 **and African ingredients** Stanley Crouch, *Considering Genius: Writings on Jazz* (New York: Basic Books, 2009), 211.

21 **different sense of time** James Jones, telephone interview by author, May 6, 2016.

22 **"past" and "present"** John S. Mbiti, *African Religions & Philosophy* (Portsmouth, NH: Heinemann, 1990), 21.

22 *Invisible Man* Ralph Ellison, *Invisible Man* (New York: Vintage, 1980), 8.

22 **delayed and compressed** Benadon, 6.

23 **"my sense of time"** David Eagleman, telephone interview by author, April 25, 2016.

25 **tenth of a second** Rooney, telephone interview by author.

25 **fee of £4** Rooney, *Ruth Belville*, 62.

25 **fifty clients** David Rooney, "Maria and Ruth Belville: Competition for Greenwich Time Supply," *Antiquarian Horology* 29, no. 5 (2006): 624.

25 **to a low setting** "Gas Lamp Danger: Inquest Warning," *Nottingham Evening Post* (Nottingham, UK), December 13, 1943, 1.

Chapter 2: Connect

27 **Stores emptied** Victor Searcher, *The Farewell to Lincoln* (Nashville, TN: Abingdon Press, 1965), 97.

28 **sojourned from Washington** John Carroll Power, *Abraham Lincoln: His Life, Public Services, Death and Great Funeral Cortege, with a History and Description of the National Lincoln Monument, with an Appendix* (Springfield, IL: E. A. Wilson & Co., 1873), 120.

28 **"appropriate transportation"** Power, 26.

29 **twelve bodies deep** Power, 132.

31 **recipes for steelmaking** Henry Bessemer, *Sir Henry Bessemer, F.R.S.: An Autobiography* (London: Offices of "Engineering," 1905), 136.

31 **gift for his sister** Bessemer, 54.

32 **tools instead of toys** R.H. Thurston, "Sir Henry Bessemer: A Biographical Sketch," *Cassier's Magazine*, September 1896, 325.

32 **He was talkative** S. T. Wellman, "The Story of a Visit to Sir Henry Bessemer: Recollection of the Early History of the Basic Open-Hearth Process," *Scientific American: Supplement*, 402.

33 **sad and preoccupied** T. J. Lodge, "A Bessemer Miscellany," in *Sir Henry Bessemer: Father of the Steel Industry*, ed. Colin Bodsworth (London: IOM Communications, 1998): 142.

34 **"many modifications"** Bessemer, *Sir Henry Bessemer, F.R.S.*, 139.

34 **"Few persons"** Bessemer, 304.

34 **While convalescing** Thurston, "Sir Henry Bessemer," 329.

34 **"I became convinced"** Bessemer, *Sir Henry Bessemer, F.R.S*, 142.

35 **"to remove the excess carbon"** Thomas J. Misa, *A Nation of Steel* (Baltimore: Johns Hopkins University Press, 1998), 8.

35 **"about ten minutes"** Bessemer, 143.

35 **"white flame"** Bessemer, 144.

36 **He entered the clothing business** Robert B. Gordon, "The "Kelly" Converter," *Technology and Culture* (1992), 769.

37 **Neither brother had experience** Gordon, 770.

37 **"fuel would be unnecessary"** J. E. Kleber and Kentucky Bicentennial Commission, *The Kentucky Encyclopedia* (Lexington: University Press of Kentucky, 1992), 485.

38 **defeated Bessemer's application** Gordon, "Kelly Converter," 777.

39 **overlooked the inconsistencies** Gordon, 778.

39 **"without the use of fuel"** William Kelly. Improvements in the Manufacture of Iron. US Patent 17,628, issued June 23, 1857.

39 **no reference to it in his letters** Gordon, "Kelly Converter," 777.

40 **steel rails lasted eighteen years** Douglas A. Fisher, *The Epic of Steel* (New York: Harper & Row, 1963), 123; Elting E. Morison, *Men, Machines, and Modern Times* (Cambridge, MA: MIT Press, 1968), 123.

40 **"smoke, lames, and sparks"** Stewart H. Holbrook, *Iron Brew* (New York: Macmillan Co., 1939), 2.

41 **"journey to New York"** Edmund Quincy, *Life of Josiah Quincy* (Boston: Little, Brown, 1874), 47.

42 *Atlas of the Geography* Charles O. Paullin, *Atlas of the Historical Geography of the United States* (Washington, DC: Carnegie Institution of Washington, 1932), 138A-B.

42 **see their grandchildren** Thomas C. Cochran, "The Social Impact of the Railroad," in *The Railroad and the Space Program*, ed. Bruce Mazlish (Cambridge, MA: MIT Press, 1965), 169.

42 **"His holy prophets"** Frank W. Blackmar, *Kansas; A Cyclopedia of State History, Embracing Events, Institutions, Industries, Counties, Cities, Towns, Prominent Persons, Etc. … With a Supplementary Volume Devoted to Selected Personal History and Reminiscence*, vol. 2 (Chicago: Standard Publishing Company, 1912), 536.

43 **distance around the equator** Ruth S. Cowan, *A Social History of American Technology* (Oxford: Oxford University Press, 1997), 117.

43 **around the world ten times** Fisher, *The Epic of Steel*, 125.

44 **wait over three centuries** Bruce David Forbes, *Christmas: A Candid History* (Berkeley: Univ. of California Press, 2008), 17.

44 **"made more of New Year's"** "Society Out Shopping," *New York Times*, December 25, 1894, 19.

44 **factories were shut** Susan G. Davis, "'Making Night Hideous': Christmas Revelry and Public Order in Nineteenth-Century Philadelphia," *American Quarterly* 34, no. 2 (1982): 187.

45 **"family dinner"** Davis.

45 **O Little Town of Bethlehem** Steven Dutch, "Making the Modern World" (lecture, University of Wisconsin-Green Bay, 2014).

45 **keep the economy going** Penne L. Restad, "Christmas in 19th-Century America," *History Today* 45, no. 12 (1995): 17.

45 **"mercantile exchanges"** "Forest of Christmas Trees," *New York Times*, December 17, 1893, 17.

45 **"a rare thing"** Restad, "Christmas in 19th-Century America," 16.

46 **"epidemic of giving"** "Heavy Christmas Mails," *New York Times*, December 21, 1890, 20.

46 **"costliness of their presents"** "Home-Made Christmas Presents," *New York Times*, December 24, 1880, 4; Forbes, *Christmas*, 116.

46 **more time to shop** Forbes, *Christmas*, 127.

47 **"greatest of human industries"** R. H. Thurston, "The Age of Steel," *Science* 3, no. 73 (1884): 792.

Chapter 3: Connect

49 **a few pirates** Daniel Walker Howe, *What Hath God Wrought: The Transformation of America, 1815–1848* (Oxford: Oxford University Press, 2007), 8.

49 **four thousand ill-prepared men** Robert V. Remini, *The Life of Andrew Jackson* (New York: Penguin, 1988), 92.

49 **trained ten thousand** Donald R. Hickey, *The War of 1812: A Forgotten Conflict, Bicentennial Edition* (Champaign: University of Illinois Press, 2012), 208.

51 **"extraordinary blunder"** Robin Reilly, *The British at the Gates: The New Orleans Campaign in the War of 1812* (New York: Putnam, 1974), 296.

52 **equal to whites** Andrew Jackson, "Proclamation: To the Free Colored Inhabitants of Louisiana," *Niles' Weekly Register*, December 3, 1814, 205. *NOHC*

53 **"bleeds afresh daily"** Letter from Samuel Morse (SFBM) to his brother Sidney Morse dated January 6, 1839, *SFBM-TWO*, 115.

54 **"intellectual branch of the art"** Letter from SFBM to his parents dated May 2, 1814, *SFBM-ONE*, 132.

55 **"intensity of attachment"** Letter from SFBM to a friend a month after the death of his wife, *SFBM-ONE*, 268.

55 **"I long to hear from you"** Letter from SFBM to his wife Lucretia Morse dated Feb 10, 1825, *SFBM-ONE*, 264.

55 **"My affectionately-beloved son"** Letter from Jedidiah Morse to SFBM dated February 8, 1825, *SFBM-ONE*, 265.

55 **Philadelphia by Monday** A letter from SFBM to Jedidiah Morse from Baltimore February 13, 1825; Samuel Irenæus Prime, *The Life of Samuel F. B. Morse* (New York: Arno Press, 1974), 144.

56 **"I wish that in an instant"** Letter from SFBM to his parents dated August 17, 1811, *SFBM-ONE*, 41.

57 **no observable difference** Prime, *The Life of Samuel F. B. Morse*, 252.

57 **"Why can't we?"** Deposition of Charles T. Jackson, Box 1, Folder 1 *SFBM-YUL*.

57 **"slight blow across the arms"** Letter dated February 1801 from SFBM to parents, *SFBM-ONE*, 19.

57 **he asked Charles T. Jackson questions** Deposition of Charles T. Jackson.

58 **"she abandoned me"** Letter from SFBM to James Fenimore Cooper dated November 20, 1849, *SFBM-TWO*, 31.

59 **brother's fireplace grate** Quote from Morse's sister-in-law, undated, *SFBM-TWO*, 21.

60 **lined in a row** Letter from SFBM to his parents dated April 1825, *SFBM-TWO*, 41.

61 **"Our democratic institutions are suffering"** Letter from SFBM on April 6, 1836 to members of the Native American Democratic Association; Carleton Mabee, *The American Leonardo: A Life of Samuel F.B. Morse* (New York: Alfred A. Knopf, 1943), 170.

61 **divine arrangement of society** Kenneth Silverman, *Lightning Man: The Accursed Life of Samuel F. B. Morse* (New York: Knopf Doubleday Publishing Group, 2010), 399.

61 **"two things well at the same time"** Letter from Jedidiah Morse to SFBM dated February 21, 1801, *SFBM-ONE*, 4.

62 **he was a coinventor** Silverman, *Lightning Man*, 156.

62 **one-third of a mile of copper** Prime, *The Life of Samuel F. B. Morse*, 303.

63 **Vail successfully conducted tests** Silverman, *Lightning Man*, 165.

63 **ten words transcribed** Stephen Vail, "The Electro-Magnetic Telegraph," *Self Culture*, May 1899, 281.

65 **"I'm still waiting, waiting"** Letter from SFBM to Sidney Morse dated January 25, 1843, *SFBM-TWO*, 191.

65 **"tantalizing and painful"** Letter from SFBM to Sidney Morse dated January 30, 1843, *SFBM-TWO*, 192.

65 **it was ridiculed** Letter from SFBM to Sidney Morse, dated January 30, 1843, *SFBM-TWO*, 193.

66 **"fraction of a dollar"** Letter from SFBM to Alfred Vail dated February 23, 1843, *SFBM-TWO*, 197.

67 **There were more hiccups** Silverman, *Lightning Man*, 226; Letter from SFBM to Sidney Morse dated January 20, 1844, *SFBM-TWO*, 216.

67 **a trickster politician** Kenneth Silverman, interviewed by Brian Lamb, *Booknotes*, C-Span, December 20, 2003, https://www.c-span.org/video/?179914-1/lightning-man-samuel-fb-morse.

69 **skin of his right arm** Candice Millard, *Destiny of the Republic* (New York: Knopf Doubleday Publishing Group, 2011), 182.

70 **"he has been murdered"** Ira Rutkow and Arthur M. Schlesinger, *James A. Garfield: The American Presidents Series: The 20th President, 1881* (New York: Henry Holt, 2006), 2.

70 **"sends his love to you"** "A Great Nation in Grief," *New York Times*, July 3, 1881, 1.

71 **a jaundiced complexion** "A Great Nation in Grief."

71 **"political necessity"** Theodore Clarke Smith, *The Life and Letters of James Abram Garfield: 1877–1882*, Vol. 2 (Hamden: Archon Books, 1968), 1184.

71 **"everything would go better"** J.C. Clark, *The Murder of James A. Garfield: The President's Last Days and the Trial and Execution of His Assassin* (Jefferson, NC: McFarland & Co., 1993), 132.

72 **equality of freed slaves** Millard, *Destiny of the Republic*, 182.

72 **came up from nothing** Millard.

72 **education and in enterprise** Millard.

72 **"a more hopeful feeling prevails"** *Complete medical record of President Garfield's case, containing all of the official bulletins, from the date of the shooting to the day of his death, together with the official autopsy, made September 20, 1881, and a diagram showing the course taken by the ball* (Washington, DC: Charles A. Wimer, 1881), 6.

73 **pleasant nap** *Complete medical record of President Garfield's case*, 11, 32, 34.

73 **bullet in the president's torso** Millard, *Destiny of the Republic*, 213.

73 **"feeling better"** *Complete medical record of President Garfield's case*, 35.

73 **"slept sweetly"** *Complete medical record of President Garfield's case*, 37.

74 **"dished up to them in this way"** Smith, *The Life and Letters of James Abram Garfield*, 1191.

74 **"telegrams from all parts"** *New York Times*, July 3, 1881, 1.

74 **"earnestly considered"** *Complete medical record of President Garfield's case*, 65.

74 **"more horrible than battle"** Millard, *Destiny of the Republic*, 215.

74 **"well-defined details"** Smith, *The Life and Letters of James Abram Garfield*, 1193.

74 **half his heft at 130** Millard, *Destiny of the Republic*, 217.

75 **cough was less** *Complete medical record of President Garfield's case*, 86.

75 **"quite weak"** *Complete medical record of President Garfield's case*, 92.

75 **"died at 10:35 p.m."** *Complete medical record of President Garfield's case*, 93.

75 **"conquered the whole"** Smith, *The Life and Letters of James Abram Garfield*, 1198.

75 **"have a place in human history?"** Millard, *Destiny of the Republic*, 228.

75 **"human hearts"** Millard.

75 **"General Garfield died"** "Trial of Guiteau," *Watchman and Southron*, November 22, 1881, 2.

76 **"one neighborhood"** U.S. Congress, House, *Electro-Magnetic Telegraphs*, HR 713, 25th Cong., 2nd sess., introduced in House April 6, 1838, House Report 753, 9. (Italics are Morse's)

76 **"Have you any news?"** David Hochfelder, *The Telegraph in America, 1832–1920* (Baltimore: Johns Hopkins University Press, 2012), 83.

77 **"'fraid of nothing"** Mary V. Dearborn, *Ernest Hemingway: A Biography* (New York: Knopf, 2017), 22.

77 **children, work, and death** Dearborn, 35.

77 **and in emergency rooms** Dearborn, 46.

77 **and thieves** Dearborn, 47.

77 **"snatched out"** Dearborn, 49.

78 **"business of writing"** Dearborn, 47.

78 **"form of storytelling"** Dearborn, 48.

78 **"Watch out for trite phrases"** "The Star Copy Style," Kansas City Star, https://www.kansascity.com/entertainment/books/article10632716.html.

78 **"Condense your language"** Letter from SMFB to Vail dated May 29, 1844, Box 1A *AV-SI.*

79 **fifteen minutes** Menahem Blondheim, *News over the Wires: The Telegraph and the Flow of Public Information in America, 1844–1897* (Cambridge, MA: Harvard University Press, 1994), 63.

79 **50 cents ($15)** Alfred Vail, *The Telegraph Register of the Electro-Magnetic Companies* (Washington, DC: John T. Towers, 1849), 10, Box 9, Folder 9 *AV-SI.*

79 **twelve words** Statistical Notebook of the Western Union Telegraph Company, Box 267, Folder 4 *WUTC-SI.*

80 **90 percent of revenue** Letter to William F. Vilas, the Postmaster General, from Norvin Green, the President of Western Union, dated November 17, 1887, Box 204, Folder 1 *WUTC-SI.*

80 **Only 2 percent** Letter to William F. Vilas.

81 *b* **for "be"** Hochfelder, *The Telegraph in America,* 75–76.

81 **"Condense your information"** Letter from SFBM to Vail dated May 25, 1844, Box 1A *AV-SI.*

81 **"best regards"** Hochfelder, *The Telegraph in America,* 75.

81 **w. 899 meant "wishful"** Francis Ormond Jonathan Smith, *The Secret Corresponding Vocabulary: Adapted for Use to Morse's Electro-Magnetic Telegraph.* (Portland, ME: Thurston, Ilsley & Co., 1845), W1000, Box 9, Folder 7 *AV-SI.*

82 *Democratic Review* "Influence of the Telegraph upon Literature," *United States Magazine and Democratic Review* 22, no. 119 (1848): 409–413.

83 **"I am somewhat disappointed"** Letter from SFBM to Vail dated August 8, 1844, Box 1A *AV-SI.*

83 **"something scary happening now"** Naomi Baron, Skype interview by the author, December 13, 2017.

84 **was a big mistake** Julia Carrie Wong, "Former Facebook Executive: Social Media Is Ripping Society Apart," *Guardian*, December 12, 2017, https://www.theguardian.com/technology/2017/dec/11/facebook-former -executive-ripping-society-apart.

Chapter 4: Capture

88 **"unsupported transit"** *Eadweard Muybridge: The Stanford Years, 1872– 1882* (Stanford, CA: Department of Art, Stanford University, 1972), 8.

89 **churning forty feet** "The Stride of a Trotting Horse." *Pacific Rural Press* (San Francisco, CA), June 22, 1878, 393.

90 **faint and fuzzy** "Quick Work," Daily Alta California, April 7, 1873, 1, *MUY-SUL*.

90 **out of a cigar box** *Eadweard Muybridge,*131.

91 **"in better relief"** E. J. Muybridge. Method and Apparatus for Photographing Objects in Motion. US Patent 212,865, filed June 27, 1878, and issued March 4, 1879.

92 **an electromagnet yanked** *Eadweard Muybridge*, 22.

92 **wires breast-high** Eadweard Muybridge, *Animals in Motion* (New York: Dover Publications, 1957), 21.

92 **"jumping from the ground"** Letter from Sherman Blake to Walter Miles dated May 6, 1929, Box 1, Folder 5, *MUY-SUL*.

93 **orange-brown blotches** Clipping "Newark Clergyman Invented Camera Film" dated March 30, 1932, *HG-NJHS-VF*.

95 **chased by a black bear** David Smith, email message to the author, November 7, 2016.

96 **"altogether new combinations"** "Kodak Film Invented Here," *Newark Sunday Call,* September 11, 1898, 1, Box 1, Folder 13, *HG-NJHS-PELL*.

97 **"transparent as glass"** F. C. Beach, "A New Transparent Film," *Anthony's Photographic Bulletin* 19, no. 5, 144.

97 **eventually fifty feet** "Goodwin's Statement," 30, Box 3, Folder 13, *HG-NJHS-PELL*.

98 **seventeen-foot long** James Terry White, *The National Cyclopedia of American Biography* (New York: J. T. White & Co., 1893), 378; Samples have been claimed to be sent to Eastman in the typed memo written by Drake & Co. dated October 26,1898, in Box 4, Folder 12 *HG-NJHS-PELL*.

98 **"settled and ended"** "Kodak Film Invented Here."

99 **"not an atom of camphor"** "Goodwin's Statement," 4.

99 **"spottedness"** "Goodwin's Statement," 22.

99 **"but much poorer"** "Goodwin's Statement," 32.

99 **they were denied** George E. Helmke, *Hannibal Goodwin and the Invention of a Base for Rollfilm* (North Plainfield: Fleetwood Museum of Art and Photographica, 1990), 5.

100 **satisfactory film until 1888** Elizabeth Brayer, *George Eastman: A Biography* (Baltimore: Johns Hopkins University Press, 1996), 192.

100 **"grind out a grist"** "Kodak Film Invented Here."

100 **"Oh, yes, yes, yes"** Handwritten memo written by Russell Everett dated November 11, 1898, Box 1, Folder 13, *HG-NJHS-PELL*.

101 **"broken in health and spirit"** Letter to Charles Pell from Rebecca Goodwin dated January 13, 1901, Box 1, Folder 15, *HG-NJHS-PELL*.

101 **the class photo** Lorna Roth, "Looking at Shirley, the Ultimate Norm: Colour Balance, Image Technologies, and Cognitive Equity," *Canadian Journal of Communication* 34, no. 1 (2009):117.

102 **lost features of their faces** Roth, 119.

102 **"film wasn't calibrated"** Oliver Chanarin, Skype interview by author, January 13, 2017.

104 **"in subtle ways"** Chanarin.

104 **"shillings per week"** John Stauffer, Zoe Trodd, and Celeste-Marie Bernier, *Picturing Frederick Douglass: An Illustrated Biography of the Nineteenth Century's Most Photographed American* (New York: Liveright Publishing, 2015), 127.

105 **most photographed human** Stauffer, Trodd, and Bernier, viii.

105 **his features were European** Marcy J. Dinius, *The Camera and the Press: American Visual and Print Culture in the Age of the Daguerreotype* (Philadelphia: University of Pennsylvania Press, 2012), 227.

106 **"a horrible botch"** W. E. B. Du Bois, "Photography," *Crisis*, October 1923, 247; Henry Louis Gates, Jr., Epilogue, *Picturing Frederick Douglass*, 198.

106 **Reports in the twentieth-century** Roth, "Looking at Shirley," 120.

106 **and chocolate manufacturers** Roth, 119.

107 **the chocolate they photographed** Roth, 120.

107 **"dark horse in low light"** Roth, 122.

108 **"Mr. V"** This section is based on interviews with Caroline Hunter. Telephone interview by author, February 21, 2017 and October 30, 2014; Interview with author in Cambridge, MA, April 13, 2017.

111 **"bad place for black people"** Interview, October 30, 2014.

112 **"desist from collaborating with the government"** Letter from G. R. Dicker to T. J. Brown dated November 9, 1970, Box i77, Folder 1–3, *POL-HBS*.

112 **changing with labor needs** Brian Lapping, *Apartheid: A History* (New York: G. Braziller, 1987), 12, 26, 77.

113 **350 passbook centers** Confidential Call Report from Hans J. Jensen to T. H. Wyman dated November 4, 1970, Box i77, Folder 1/3, *POL-HBS*.

114 **"no company, no investment"** Edwin Land, Polaroid Shareholder Meetings 1971, audio tape recording, *POL-HBS*.

114 **Polarizer South Africa** Handwritten and undated note titled "Chronology," 1977, Box 79, Folder 2/3, *POL-HBS*.

115 **only sixty-five ID-2 cameras** Polaroid Memo from G. R. Dicker to All Polaroid Employees, dated October 6, 1970, Box 77, Folder 1/2, *POL-HBS*.

115 **"tissue of lies"** Christopher Nteta, speech given on October 8, 1970, Box 80, *POL-HBS*.

115 **liberation movements** Polaroid Memo dated October 7, 1970, from the Polaroid Revolutionary Workers Movement to Edwin Land, Box i78, Folder 1/2, *POL-HBS*.

116 **refused to work with South Africa** Polaroid Statement dated October 7, 1970, Box i77, Folder 1/2, *POL-HBS*.

116 **155 black employees** Confidential Call Report from Hans J. Jensen to T. H. Wyman dated November 4, 1970, Box i77, Folder 1/3, *POL-HBS*.

116 **an indoctrination of inferiority** "Polaroid Announces 'Experiment' to Help Blacks in South Africa." *Harvard Crimson*, January 14, 1971.

117 **"late at night"** Polaroid Confidential Memo from James Shea to Polaroid Management dated July 25, 1972, Box 78, Folder 2/4, *POL-HBS*.

117 **"I'm mad at them"** Polaroid Revolutionary Workers Movement Press Release dated February 11, 1971, Box 2, Folder 1, *PRWM-SCH*.

117 **"do little experiments"** Edwin Land, Polaroid Shareholder Meetings 1971.

118 **unlabeled boxes of film** Robert Lenzner, "Polaroid's S. Africa Ban Defied?," *Boston Globe*, November 21, 1971, 1977, Box 79, Folder 2/3, *POL-HBS*.

119 **"exactly 42%"** David Smith, "'Racism' of Early Colour Photography Explored in Art Exhibition," *Guardian*, January 25, 2013, https://www.theguardian.com/artanddesign/2013/jan/25/racism-colour-photography-exhibition.

Chapter 5: See

122 **number is decreasing** Sara Lewis, Skype interview by author, February 21, 2017; James Lloyd, telephone interview by author, March 10, 2017.

122 **"every six months"** Robert Friedel, "New Lights on Edison's Light," *Invention & Technology* 1, no. 1: 24.

123 **to see a new electrical invention** Robert Friedel, Paul Israel, and Bernard S. Finn, *Edison's Electric Light* (New Brunswick, New Jersey: Rutgers University Press, 1986), 5.

123 **best physics departments** William Hammer, "William Wallace and His Contributions to the Electrical Industries (Part II)," *Electrical Engineer* XV, no. 249: 130; William Hammer, "William Wallace and His Contributions to the Electrical Industries (Part I)," *Electrical Engineer* 15, no. 248: 105.

124 **Eloise was as knowledgeable** Hammer, "William Wallace," no. 249: 129.

124 **entertain prominent scientists** Hammer, "William Wallace," no. 248: 104.

125 **his first arc light** William Hammer, "William Wallace and His Contributions to the Electrical Industries (Part III)" *Electrical Engineer* XV, no. 250: 159.

126 **two shifts** "Invention's Big Triumph," *New York Sun*, September 10, 1878, 1, *TAE-RU*.

126 **"4,000 candles"** "Invention's Big Triumph."

126 **reasonable means of illumination** Brian Bowers, *A History of Electric Light & Power* (Stevenage, UK: Peter Peregrinus Press, 1982), 8.

126 **"power may be transmitted"** "Invention's Big Triumph."

127 **"under the electric lights"** "He Showed Edison the Light," *Sunday Republican*, November 8, 1931, Features Section, 1, *DHS*.

127 **"I do not think you are working"** Hammer, "William Wallace," no. 248: 105.

128 **"brought into private houses"** Paul Israel, *Edison: A Life of Invention* (New York: Wiley, 2000), 166.

128 **England, France, Russia** Friedel, Israel, and Finn, *Edison's Electric Light*, 115.

128 **"I have it."** "Edison's Newest Marvel," *New York Sun*, September 16, 1878, 3, *TAE-RU*.

129 **half-heartedly fiddled** Matthew Josephson, *Edison: A Biography* (London: Eyre & Spottiswoode, 1961), 178.

129 **Platinum was promising** Friedel, Israel, and Finn, *Edison's Electric Light*, 16.

130 **initially rejected** Friedel, Israel, and Finn, 93.

131 **"diabetes, and obesity"** Mariana Figueiro, telephone interview by author, September 8, 2016.

132 **the function of the eye** David M. Berson, Felice A. Dunn, and Motoharu Takao, "Phototransduction by Retinal Ganglion Cells That Set the Circadian Clock," *Science* 295, no. 5557 (2002): 1070–1073.

133 **"evolved with us"** Thomas Wehr, telephone interview by author, July 14, 2016.

134 **"it is just by chance"** Richard Stevens, telephone interview by author, July 21, 2016, and October 18, 2016.

137 **can see about fifty** International Dark-Sky Association, *Fighting Light Pollution: Smart Lighting Solutions for Individuals and Communities* (Mechanicsburg, PA: Stackpole Books, 2012), viii.

137 **clouds bounce and reflect light** Christopher C. M. Kyba et al., "Cloud Coverage Acts as an Amplifier for Ecological Light Pollution in Urban Ecosystems," *PLoS ONE* 6, no. 3 (2011): e17307.

137 **"lose the contrast"** Fabio Falchi, Skype interview with author, October 18, 2016.

138 **"gray, silvery cloud"** Ron Chepesiuk, "Missing the Dark: Health Effects of Light Pollution," *Environmental Health Perspectives* 117, no. 1 (2009): A22.

138 **two out of three people** Pierantonio Cinzano, Fabio Falchi, and Christopher D Elvidge, "The First World Atlas of the Artificial Night Sky Brightness," *Monthly Notices of the Royal Astronomical Society* 328, no. 3 (2001): 689–707.

138 **"all of human history has inspired us"** Paul Bogard, telephone interview by author, August 31, 2016.

139 **"I am a *photonis greeni*"** Lewis, Skype interview by author, February 21, 2017.

140 **one molecule of ATP** Lloyd, telephone interview by author, March 10, 2017.

140 **"insects are nocturnal"** Bogard, Skype interview by author, August 31, 2016.

140 **"die this way"** Travis Longcore, Skype interview by author, September 12, 2016.

141 **shelter from predators** Michael Salmon, "Protecting Sea Turtles from Artificial Night Lighting at Florida's Oceanic Beaches," in *Ecological Consequences of Artificial Night Lighting*, ed. Catherine Rich and Travis Longcore (Washington, DC: Island Press, 2006), 148.

141 **the banisters** *The City Dark*, Directed by Ian Cheney, 83 minutes, 2011.

142 **sixty-five years old** George C. Brainard, Mark D. Rollag, and John P. Hanifin, "Photic Regulation of Melatonin in Humans: Ocular and Neural Signal Transduction," *Journal of Biological Rhythms* 12, no. 6 (1997): 542.

142 **"no link between lighting and crime"** Paul Bogard, *The End of Night: Searching for Natural Darkness in an Age of Artificial Light* (Boston: Little, Brown, 2013), 79.

143 **removing the blue light they produce** *Human and Environmental Effects of Light Emitting Diode (LED) Community Lighting.* American Medical Association Council on Science and Public Health (2016), https://www.ama-assn.org/sites/ama-assn.org/files/corp/media-browser/public/about-ama/councils/Council%20Reports/council-on-science-public-health/a16-csaph2.pdf

143 **10 percent of city streetlights** *Human and Environmental Effects of Light Emitting Diode (LED) Community Lighting.*

143 **99 percent of people** Cinzano, Falchi, and Elvidge, "First World Atlas," 689.

Chapter 6: Share

146 **artifacts of human life** A typed draft of dust cover copy for "Murmurs of Earth," Box 1247, Folder 5, *SAG-LC.*

146 **"Absolutely"** John Casani, telephone interview by author, August 23, 2018.

146 **bins at Tower Records** Timothy Ferris, telephone interview by author, August 23, 2018.

147 **strummed a human chord** Ann Druyan, "Earth's Greatest Hits," *New York Times Magazine,* September 4, 1977, 13.

147 **"vote for Bach"** Lewis Thomas, *Lives of a Cell* (London: Bantam, 1974), 45.

147 **playlist began to reflect** Bertram Lyons, telephone interview by author, July 13, 2018; "Alan Lomax and the Voyager Golden Records," https://blogs.loc.gov/folklife/2014/01/alan-lomax-and-the-voyager-golden-records/

149 **while working on both the telephone** Robert A. Rosenberg et al., *The Papers of Thomas A. Edison: Menlo Park: The Early Years, April 1876–December 1877,* vol. 3 (Baltimore, MD: Johns Hopkins University Press, 1989), 444. (This volume will be called *TAE* hereafter.)

149 **were inaudible** Rosenberg et al., 472.

150 **"wax paper underneath"** Rosenberg et al., 699.

150 **a circular diaphragm** Rosenberg et al.,

150 **"a distinct sound"** George Parsons Lathrop, "Talks with Edison," *Harper's New Monthly Magazine,* February 1890, 430.

151 **first six days of December** *TAE,* 649.

151 **"taken aback in my life"** Frank Lewis Dyer and Thomas Commerford Martin, *Edison: His Life and Inventions*, vol. 1 (New York: Harper & Brothers, 1910), 208.

151 **"ary ad ell am"** *TAE*, 699.

151 **a minute of sound** Matthew Josephson, *Edison: A Biography* (London: Eyre & Spottiswoode, 1961), 173.

152 **"How do you like the phonograph?"** "The Phonograph," *Scientific American* 75, no. 4: 65; "The Talking Phonograph," *Scientific American* 37, no. 25: 384.

152 **"Speech has become"** Edward H. Johnson, "A Wonderful Invention—Speech Capable of Indefinite Repetition from Automatic Records," *Scientific American* 37, no. 20: 304.

152 **answering machines** Thomas A. Edison, "The Phonograph and Its Future," *North American Review* 126, no. 262 (1878): 533–535.

153 **sheet music played** Andre Millard, *America on Record: A History of Recorded Sound* (Cambridge, UK: Cambridge University Press, 2005), 108.

153 **"American music and musical taste"** Steven D. Lubar, *Infoculture: The Smithsonian Book of Information Age Inventions* (New York: Houghton Mifflin Harcourt Publishing Co., 1993), 173.

153 **twenty-six million records** Lubar, 174.

153 **a hundred million** Lubar, 177.

154 **Cellos, violins, and guitars . . . trombones** Millard, *America on Record*, 80–83.

155 **130 million cassette tapes** Millard, 5.

157 **sixteen billion punchcards** Rey Johnson, "The First Disk File" (Dinner Talk, DataStorage '89 Conference, San Jose, California, September 19, 1989), http://www.mdhc.scu.edu/100th/reyjohnson.htm.

158 **rider's physical features** Typed Letter from Herman Hollerith to Mr. Wilson dated August 7, 1919, Box 9, Folder 7, *HH-LC*.

158 **two months to tally** Charles W. Wootton and Barbara E. Kemmerer, "The Emergence of Mechanical Accounting in the US, 1880–1930," *Accounting Historians Journal* 34, no. 1 (2007): 105.

160 **headed to the junkyard** "RAMAC: An Everlasting Impact on the Computer Industry" by Thomas J. Watson Jr., undated, Folder 24, *RJ-CHM*.

160 **from the graveyard** "RAMAC."

160 **a Dixie cup** Lab notebook No. 22–83872, 139, dated November 10, 1953, Box 7, *JJH-WPI*.

161 **wife's old silk stocking** Typed document "Biographic Sketch," 2, Box 2, *JJH-WPI*.

161 **hard, tough, and smooth** Memo entitled "The Invention and Development Process," from Jacob Hagopian to Albert Hoaglan, dated June 18, 1975, Box 2, *JJH-WPI*.

161 **reached out to companies** "Magnetic Ink and Powders," memo from Jacob Hagopian (JH) to E. Quade, April 22, 1959, Box 8; Letter from Ferro Enameling Company, February 2, 1954, Box 2; Letter from Reeves Soundcraft, Dec 15, 1953, Box 2; Letter from W. P. Fuller Company, December 22, 1953, Box 2 *JJH-WPI*.

162 **too weak to store data** Letter from Jacob Hagopian to Reynold Johnson dated May 20, 1992, Box 2, *JJH-WPI*.

162 **"My complaint is"** Undated handwritten draft of letter by Jacob Hagopian to the editor of Newstack, Box 2, *JJH-WPI*.

Chapter 7: Discover

166 **but nothing spectacular** John Drury Ratcliff, *Yellow Magic: The Story of Penicillin* (New York: Random House, 1945), 13.

166 **animal figurines** Kevin Brown, interview by author, London, England, September 27, 2017.

166 **using bacteria for his pigments** Eric Lax, *The Mold in Dr. Florey's Coat: The Story of the Penicillin Miracle* (New York: Henry Holt and Co., 2005), 12.

166 **bit of a slob** Brown interview.

166 **"That's funny"** Lax, *Dr. Florey's Coat*, 17.

streptococcus, staphylococcus, *Yellow Magic*, 22.

168 **family of glassmakers** William Ernest Stephen Turner, "Otto Schott and His Work. A Memorial Lecture," *Journal of the Society of Glass Technology* 20(1936): 83.

168 **blacks, reds, and blues** Simon Garfield, *Mauve* (London: Faber & Faber, 2013), 8.

169 **melting of glass** Turner, 85.

170 **"predictable properties was required"** Jurgen Steiner, "Otto Schott and the Invention of Borosilicate Glass," *Glastechnische Berichte* 66, no. 6–7 (1993): 166.

170 **wrote a letter** Steiner.

171 **from the same column** Turner, "Otto Schott and His Work," 86.

172 **"has been solved"** Steiner, "Otto Schott," 166.

172 **there were seventy-six** Turner, "Otto Schott and His Work," 90.

173 **boron played different roles** Jane Cook, telephone interview by author, June 9, 2017 and November 1, 2017.

174 **they hired scientists** Margaret B.W. Graham and Alec T. Shuldiner, *Corning and the Craft of Innovation* (Oxford, UK: Oxford University Press, 2001), 38.

175 **That speed came at a cost** Graham and Shuldiner, 46.

176 **a precipitous drop** Graham and Shuldiner, 55.

176 **but his colleagues laughed** "The Battery Jar That Built a Business: The Story of Pyrex Ovenware and Flameware." *Gaffer*, July 1946, 3, *COR*.

176 **broke like glass** John Littleton, telephone interview by author, September 7, 2017.

176 **buried in a glass coffin** Harvey K. Littleton, interview by Joan Falconer Byrd, Spruce Pine, N.C., March 15, 2001, transcript, Archives of American Art, Smithsonian Institution, Washington, DC, https://www.aaa.si.edu/collections/interviews/oral-history-interview-harvey-k-littleton-11795.

177 **colored people at her table** Joseph C. Littleton, "Recollections of Mom: By Her Third Child, Joe," (unpublished book, 1995), 73 and 77, *CMOG*.

177 **Guernsey casserole dish** A Report of the History of the First Pyrex Baking Dish, dated November 1917, *COR*.

177 **Bessie didn't cook** Nancy Jo Drum, telephone interview by author, September 11, 2017.

178 **Using every bowl** J.C. Littleton, 16.

178 **steak to cocoa** History of the First Pyrex Baking Dish.

178 **didn't retain the flavor** "The Battery Jar That Built a Business," 3, *COR*.

178 **it contained lead** "Informal Notes as taken from Dr. Sullivan: History of Pyrex bakeware," dated March 5, 1954, *COR*.

179 **dropped waist high** E. C. Sullivan, "The Development of Low Expansion Glasses," *Journal of the Society of Chemical Industry* 15, no. 9 (1916): 514.

179 **silver coating did not cook well** Sullivan.

179 **called "Py-right"** Graham and Shuldiner, *Corning and the Craft of Innovation*, 56.

179 **renamed to Pyrex in 1915** "The Battery Jar That Built a Business," 5, *COR*.

179 **4.5 million pieces** "The Battery Jar That Built a Business," 6.

180 **President Woodrow Wilson** Daniel Kelm, telephone interview by author, September 18, 2017.

180 **aerial cameras and rangefinders** Edward J. Duveen, "Key Industries and Imperial Resources," *Journal of the Royal Society of Arts* 67, no. 3459 (1919): 242.

180 **thousands of German patents** "The 'Trading with the Enemy Act'" *Scientific American* 117, no. 20 (1917): 363.

181 **and pressure gage covers** W. H. Curtiss, "Pyrex: A Triumph for Chemical Research in Industry," *Industrial & Engineering Chemistry* 14, no. 4 (1922): 336–337.

181 **Huge tariffs were placed** Graham and Shuldiner, *Corning and the Craft of Innovation*, 59.

183 **a botanist** J.J. Thomson, *Recollections and Reflections* (New York: The Macmillan Co., 1937), 6.

184 **"nor wholly wrong"** Thomson, 376.

184 **Victorian bull** D. J. Price, "Sir J.J. Thomson, OM, FRS. A Centenary Biography." *Discovery* 17: 496, *JJT-TRI*.

185 **hammer in the house** George Paget Thomson, "J.J. Thomson and the Discovery of the Electron," *Physics Today* 9, no. 8 (1956): 23.

185 **"more easily said than done"** J.J. Thomson, *Recollections and Reflections*, 334.

186 **"glass in the place is bewitched"** Baron Robert John Strutt Rayleigh, *The Life of Sir JJ Thomson, Sometime Master of Trinity College, Cambridge* (London: Dawsons, 1969), 25.

186 **In the summer of 1897** Isobel Falconer, "JJ Thomson and the Discovery of the Electron," *Physics Education* 32, no. 4 (1997): 227.

187 **"no telephone, no radio"** J.J. Thomson, *Recollections and Reflections*, 1.

Chapter 8: Think

189 **sliding under his left cheek** Hanna Damasio et al., "The Return of Phineas Gage: Clues About the Brain from the Skull of a Famous Patient," *Science* 264, no. 5162 (1994): 1104.

190 **"as well as an iron frame"** John M. Harlow, "Recovery from the Passage of an Iron Bar through the Head," *History of Psychiatry* 4, no. 14 (1993): 275.

190 **affable, reliable, and clever** Damasio et al., "Return of Phineas Gage," 1102.

190 **"no longer Gage"** Harlow, "Recovery from the Passage," 274.

190 **distracted, unreliable, fitful** Damasio et al., "Return of Phineas Gage," 1104.

191 **Stone Age brains** Torkel Klingberg, *The Overflowing Brain: Information Overload and the Limits of Working Memory* (Oxford, UK: Oxford University Press, 2009), 3.

191 **simple tool of fire** Richard Wrangham, *Catching Fire: How Cooking Made Us Human* (New York: Basic Books, 2009), 120.

191 **enhanced by television** N.C. Andreasen, *The Creative Brain: The Science of Genius* (New York: Plume, 2006), 146.

192 **bigger than in nonmusicians** Kenneth M. Heilman, "Possible Brain Mechanisms of Creativity," *Archives of Clinical Neuropsychology* 31, no. 4 (21 March 2016): 287.

192 **London taxicab drivers** Eleanor A. Maguire et al., "Navigation-Related Structural Change in the Hippocampi of Taxi Drivers," *Proc. Natl. Acad. Sci.* 97, no. 8 (2000): 4398–4403.

192 **learn how to juggle** Klingberg, *The Overflowing Brain*, 12.

193 **"wonderful thing"** R.V. Bruce, *Bell: Alexander Graham Bell and the Conquest of Solitude* (Ithaca, NY: Cornell University Press, 1990), 198.

194 **"A more interesting performance"** "The Telephone Concert." *New Haven Evening Register*, April 28, 1877, 4, *NHFPL*.

194 **"I commenced at once"** Joseph Leigh Walsh, *Connecticut Pioneers in Telephony: The Origin and Growth of the Telephone Industry in Connecticut* (New Haven, CT: Morris F. Tyler Chapter, Telephone Pioneers of America, 1950), 327.

194 **"the greatest invention of recent years"** "The Telephone Concert," 4.

195 **"just as gas and water are supplied"** The Telephone." *New Haven Daily Morning Journal and Courier* (New Haven, CT), April 28, 1877, 2. *CSL*.

196 **habit of calling, complaining** Herman Ritterhoff, "How a Laugh Lost a Millon," *Telephony*, March 22, 1913, 59.

197 **wounded at Winchester** Kathy Kanauer, "Almon Strowger" (lecture, Penfield Historical Association, Penfield, NY, February 27, 2000), 2, *AS-PEN*.

198 **so he could use the box** J. Hartwell Jones, "Industry Honors First Automatic Inventor," *Telephony*, October 15, 1949, 12, *AS-PEN*.

198 **short-circuiting the line** Herman Ritterhoff, "How a Laugh Lost a Millon," *Telephony*, March 22, 1913, 59.

204 **"I can make you a rod of germanium"** Gordon K. Teal, "Single Crystals of Germanium and Silicon—Basic to the Transistor and Integrated Circuit," *IEEE Transactions on electron devices* 23, no. 7 (1976): 623.

205 **took apart a clock to get its motor** Michael F. Wolff, "Innovation: The R&D 'Bootleggers': Inventing against Odds," *IEEE Spectrum* 12, no. 7 (1975): 41.

205 **he inserted his pen** Pawel E. Tomaszewski and Robert W. Cahn, "Jan Czochralski and His Method of Pulling Crystals," *MRS Bulletin* 29, no. 5 (2004): 348–349.

206 **"pigheaded." "I thought that was silly"** Gordon Teal, interview by Lillian Hoddeson and Michael Riordan, June 19, 1993, transcript, The Niels Bohr Library & Archives Oral History, American Institute of Physics, College Park, MD, 11 (used with permission).

206 **Teal's day had two starts** Teal, "Single Crystals of Germanium," 625.

207 **"My family sort of felt that they lost me"** G. Teal interview, 13.

207 **he didn't talk baseball** Donald Teal, telephone interview by author, January 26, 2017.

208 **"Summit Ridge Road"** Michael Riordan, "The Lost History of the Transistor," *IEEE Spectrum* 41, no. 5 (2004): 45.

209 **"Do you have silicon transistors"** Riordan.

209 **"I just happen to have some"** John McDonald, "The Men Who Made T.I.," *Fortune*, November 1961, 226, Box 11, Folder 17, *BAY*.

209 **"They have silicon transistors in Texas!"** McDonald.

211 **"make us much, much smarter"** David Eagleman, telephone interview by author, May 7, 2018.

211 **This strains our brains** Nicholas Carr, *The Shallows: What the Internet Is Doing to Our Brains* (New York: W. W. Norton, 2011), 126.

211 **yank, tug, jerk, and jolt** Carr, 91.

211 **scanning superficially** Carr, 138.

211 **weaken our ability to think deeply** Carr, 120.

211 **scratch paper** Carr, 123.

211 **happening in the working memory** Klingberg, *The Overflowing Brain*, 130.

212 **to mathematically produce** Sheldon Hochheiser, email to the author, May 7, 2018.

212 **magic number is seven** George A. Miller, "The Magical Number Seven, Plus or Minus Two: Some Limits on Our Capacity for Processing Information," *Psychological Review* 101, no. 2 (1994): 343.

212 **pool of the long-term memory** Carr, *The Shallows*, 124.

212 **so we end up scatterbrained** Carr, 118.

213 **the 2001 movie** *Memento,* Directed by Christopher Nolan, 1h 53min, 2000.

213 **as an extended mind** Andy Clark and David Chalmers, "The Extended Mind," *Analysis* 58, no. 1 (1998): 7–19.

213 **"tools in the world"** David Chalmers, telephone interview by author, May 7, 2018.

214 **where they are located.** Betsy Sparrow, Jenny Liu, and Daniel M. Wegner, "Google Effects on Memory: Cognitive Consequences of Having Information at Our Fingertips," *Science* 333, no. 6043: 778.

214 **"where" over "what"** Sparrow, Liu, and Wegner.

214 **"It's not an important loss"** Eagleman interview.

215 **Musicians have an increase** Heilman, "Possible Brain Mechanisms," 287–288.

215 **"then break and blend"** Eagleman interview.

215 **innovation, and production** Heilman, "Possible Brain Mechanisms," 285.

215 **"The Internet can provide the investigator"** Heilman, email to author, May 2, 2018.

216 **more susceptible to being distracted** Klingberg, *The Overflowing Brain*, 73.

216 **Our hunting and gathering minds** Carr, *The Shallows*, 138.

216 **There are no computers!** Nick Bilton, "Steve Jobs Was a Low-Tech Dad," *New York Times*, September 10, 2014, E2. https://www.nytimes.com/2014/09/11/fashion/steve-jobs-apple-was-a-low-tech-parent.html.

216 **"low-tech dad"** Bilton.

216 **IQ test scores rose** Klingberg, *The Overflowing Brain*, 13.

216 **"We are doing in minutes"** Thomas Alva Edison and Dagobert D. Runes, *The Diary and Sundry Observations of Thomas Alva Edison* (Philosophical Library, 1948), 107.

Epilogue

220 **"fresh point of view"** Toni Morrison, "Address to the Second Chicago Humanities Festival, Culture Contact" (lecture, Word of Mouth Series, Chicago, IL, 1991), https://www.youtube.com/watch?v=KxqQhkMKlC0.

Annotated Bibliography

The next few pages serve as the extended notes, commentary, and suggested resources for the stories described in this book. Headings are provided so that a reader can save time and get the information they desire readily. In many instances, there are topics that have very few sources; in other cases, there are some with too many. Where there is an abundance of material, the key resources to use are provided, as well as the pool from which those selections are taken. It is hoped that the serious reader or scholar will be able to get from zero to hero on a particular topic in a very short amount of time. Happy hunting!

Chapter 1: Interact

Ruth Belville. The story of Ruth Belville has best been captured in the succinct book called *Ruth Belville: The Greenwich Time Lady* by David Rooney, who painstakingly put together pieces about the life of this woman who sold time. Ruth Belville is also briefly mentioned in older books, which include: *British Time* (1947) by Donald de Carle and *Greenwich Time and the Discovery of the Longitude*

(1980) by Derek Howse. Both of these books are worth the hunt for any reader with a burgeoning interest in timekeeping. They surpass many other books, including more recent titles. As for additional sources, British newspapers have a few references about her. Belville was a bit of a celebrity and there are several articles mentioning her, particularly around the time of her death in 1943. As for contemporary accounts about Ruth Belville, there are a few scholarly works to consider. These include "Clock Synchrony, Time Distribution and Electrical Timekeeping in Britain 1880–1925" by Hannah Gay, and "Maria and Ruth Belville: Competition for Greenwich Time Supply" by David Rooney. Both of these works are dense with details, like the cost of Belville's services, and Rooney's paper cites Ruth Belville's letters, rendering itself a treasure trove for any eager reader wanting more about this entrepreneurial woman. For short summaries about Ruth Belville, two magazine articles will serve as great introductions to her: John Hunt's "The Handlers of Time" and Stephen Battersby's "The Lady Who Sold Time."

Sleeping Patterns. Sleep is a topic that has become a national concern. We dwell in what the *New York Times* calls the "sleep-industrial complex," where drug companies and mattress manufacturers are making billions of dollars from our sleep anxiety. Newspaper articles, magazine articles, and websites have begun to discuss this topic; however, scholarly work on the shifting patterns of our sleep, such as *At Day's Close: Night in Times Past* by A. Roger Ekirch and *The Slumbering Masses: Sleep, Medicine, and Modern American Life* by Matthew Wolf-Meyer, will awaken one's understanding of how we got here. For reliable medical information on the sleeping patterns of Americans, the National Institute of Health's (NIH) website contains reports and statistics that will quench any thirst. Also, the website for the Centers for Disease Control and Prevention (CDC) stores data and charts about sleeping pill consumption in the United States. Lastly, the top scientists in the land have put together a call-to-action in a report entitled "Extent and Health Consequences of Chronic Sleep Loss and Sleep Disorders," which is downloadable from the National Academies Press's website.

Benjamin Huntsman. Readers interested in the life of Benjamin Huntsman are fortunate, for a short and understandable text about him exists. The Sheffield City Libraries (in the UK) published a ten-page booklet written by scholar Kenneth C. Barraclough called *Benjamin Huntsman 1704–1775* (available for purchase through the Sheffield Library). This pamphlet documents everything from his birth to his death, and the interesting (and provocative) life he had in between, such as his forbidden marriage and divorce. These particulars were omitted from this book because they did not serve the story and would have distracted the reader, but the inquisitive should seek out this little gem. In addition to Barraclough's brief account of Huntsman, there are thick older books like *Industrial Biography: Iron Workers and Tool Makers* (1863) by Samuel Smiles, which profiles Huntsman and other contributors to the manufacture of iron and steel. For additional information about Huntsman's life and lineage, the articles "The Pedigree and Career of Benjamin Huntsman" by W. Wyndham Hulme, and "Benjamin Huntsman, of Sheffield, the Inventor of Crucible Steel" by R. A. Hadfield are essential works. For highly technical and scientific accounts of Huntsman's creation, the prolific Kenneth C. Barraclough wrote *Steelmaking Before Bessemer: Volume 1—Blister Steel*. This is an authoritative but hard-to-acquire book. Any serious scholar who desires more about the history and science of metals should also pick up Cyril Smith's *A Search for Structure* or R. F. Tylecote's *A History of Metallurgy*.

Galileo. Galileo's story has been known for centuries, but it remains fresh and interesting for every generation. He has been the focus of a few popular titles, such as Dava Sobel's *Galileo's Daughter: A*

Historical Memoir of Science, Faith and Love and the earlier *Galileo: Pioneer Scientist* by Stillman Drake. Galileo was an astronomer and physicist famous for his discovery of Jupiter's moons; but his experiments at the Tower of Pisa and his work on the pendulum swing both have an impact on our daily lives. While many students can give details about his dropping of different objects, most do not know about Galileo's pendulum clock. For details on his efforts to create a clock, *The Pulse of Time* by Silvio Bedini is a rare, well-researched tome that is a must-have for the highly inquisitive or any serious scholar. Interestingly, there is still a debate among academicians as to whether this tale of the swinging lamp in the church is true. Be that as it may, what is certain is that discussion of Galileo and the challenge to mark off the hours remains interesting—and timeless.

Warren Marrison. For a scientist who made such a significant impact on society, very little has been written about Warren Marrison. This book aimed to correct that oversight, but earlier efforts are essential works in the study of this inventor. A short biography called "Warren A. Marrison—Pioneer of the Quartz Revolution" was written in 1989 by W. R. Topham and can be acquired from the National Association of Watch and Clock Collectors. Bell Laboratories, Marrison's former employer, has a short entry about him in the first volume of the very thick two-volume set called *A History of Engineering and Science in the Bell System: The Early Years (1875–1925)*. Marrison's work is mentioned on pages 319 and 991. But that's it! Fortunately, Marrison took matters of preserving his legacy into his own hands and summarized his contribution to timekeeping in his long article called "The Evolution of the Quartz Crystal Clock."

Piezoelectricity. Piezoelectricity is such a fascinating materials-science phenomenon, one would expect there to be more accessible material written about it. That is not the case, however. The most seminal work is Walter Cady's book *Piezoelectricity,* and the first chapter of this book will serve the curious reader, but Cady's text nosedives into highly technical discussions after just a few pages. For a discussion with a less steep drop, there are several introductory materials-science books, particularly ones on the topics of smart materials or ceramics that might be useful. As for the discovery of piezoelectricity, there is a biography about Pierre Curie written by his famous wife, Marie Curie. For those interested in the early history and uses of piezoelectricity, the multiple research articles by Shaul Katzir will certainly provide many items of interest.

The Impact of Timekeeping. Just as the word *time* is one of the most used words in the English language, there is no dearth of books on the concepts of time and timekeeping. There are a few that stand out, however: Carlene Stephens has written a fun, readable, illustrated text on the evolution of timekeeping and its impact on society called *On Time: How America Has Learned to Live Life by the Clock.* Also, the book *Revolution in Time: Clocks and the Making of the Modern World* by David Landes is a seminal, scholarly work that is a standard text on the topic of timekeeping. Books like *The Geography of Time: The Temporal Misadventures of a Social Psychologist* (Robert Levine) and *Hidden Rhythms* (Eviatar Zerubavel) provide a thorough discussion of how life changed by way of the clock, but there are certainly plenty of other books on this topic that will serve the reader as well. For an interesting treatment on the cultural differences in the perception of time, articles such as James Jones's "Cultural and Individual Differences in Temporal Orientation" and Robert Levine's and Ara Norenzayan's "The Pace of Life in 31 Countries," are very informative. Jones's piece is particularly useful in the unpacking—and supporting—of the notion of CP (colored people's) time.

Chapter 2: Connect

Lincoln's Funeral Train. The funeral procession for Abraham Lincoln was once part of America's collective memory, but over time this event has slipped from the nation's consciousness. For readers seeking rich descriptions about Lincoln's funeral train, Victor Searcher's *The Farewell to Lincoln* is the principal text to read. There is also a self-published book called *The Lincoln Train Is Coming* by Wayne and Mary Cay Wesolowski, which is an invaluable resource but hard to come by. This title is well worth the hunt, for it contains vivid descriptions and facts culled from numerous newspapers, as well as the authors' own research. (For example, the color of the funeral car was a mystery until Wayne Wesolowski found a piece and analyzed it.) Many of the books about Lincoln's procession are a few decades old. However, there has been a succession of more contemporary tellings of this great national tale, published in time for the 150th anniversary. One such book, released in 2014, is called *Lincoln's Funeral Train: The Epic Journey from Washington to Springfield* by Robert M. Reed. There is also an illustrated children's book called *Abraham Lincoln Comes Home* by Robert Burleigh, which tenderly memorializes this great moment in history for young people. Additional accounts about Lincoln's cortege can also be found in scores of newspapers from the various cities that the train traveled through or where processions were held.

Sir Henry Bessemer. Sir Henry Bessemer was the father of the steel industry, but he never received the authoritative biography he deserved. He attempted to correct this by writing one himself, which he called *Sir Henry Bessemer, F. R. S.: An Autobiography*. Despite being overlooked during his lifetime, there have been modern efforts to fix this historical deficiency. In the twentieth century, the Institute of Materials published *Sir Henry Bessemer: Father of the Steel Industry*, which not only describes the man, but the business he inspired.

This text is technical in part, but there are anecdotal accounts about Bessemer from some of the last people who knew him. Additionally, the Herne Hill Society in the UK published a small book called *The Story of Sir Henry Bessemer*, but this book is difficult to acquire in the United States. Most of the particulars about Bessemer's life are limited to the tales he was willing to share in his autobiography and from the newspaper articles about him. While steel is abundant, we know less about the man than about the metal he created. For an authoritative and technical tome about Bessemer's invention, Kenneth C. Barraclough's *Steelmaking, 1850–1900* is an essential text to read.

William Kelly. While there is a paucity of accounts about Sir Henry Bessemer, the pickings are even slimmer for William Kelly. The books *Iron Brew: A Century of American Ore and Steel* by H. Holbrook Stewart and *Men, Machines, and Modern Times* by Elting E. Morison discuss Kelly to some extent. Morison spends the majority of his book on the invention of steel, describing it in a nearly narrative style. This book is a great addition for any steel aficionado. For additional information about William Kelly, the best bets are encyclopedias as well as nineteenth-century books and newspaper articles. In a few of these old accounts, letters written by Kelly are referred to, but they seem to have been lost. Despite these missing artifacts, there is a bigger issue with Kelly that requires addressing: in particular, Kelly's claim that he brought steel to the United States is largely manufactured out of whole cloth. The most definitive and well-researched study on Kelly's steelmaking efforts is the academic paper, "The 'Kelly' Converter," by Robert Gordon, a professor emeritus of Yale University who is an expert in archaeometallurgy. Professor Gordon spent time with Kelly's converter at the Smithsonian and analyzed its materials. His research shows that Kelly's contribution to steelmaking lacks proof. (Landmarks and plaques continue to be installed in Kelly's honor, despite these learned observations.)

Steel and Rails. For a material so important to society, there are very few recent books about steel written for lay audiences. One title that stands out is an older text called *Metals in the Service of Man* by Arthur Street and William Alexander, which discusses steel in a few of its chapters in a very easy-to-read style. Additionally, a chapter in Stephen Sass's *The Substance of Civilization: Materials and Human History from the Stone Age to the Age of Silicon* provides a solid survey about the science behind steel. For an examination of the role of steel on culture, a few older books are serviceable, such as *The Epic of Steel* by Douglas Alan Fisher (and to a lesser extent his other book *Steel Serves the Nation*) and *The Coming of Age of Steel* by Theodore A. Wertime. The history of metallurgy going back to ancient times can be found in R. F. Tylecote's *A History of Metallurgy*. For modern studies about steelmaking, Thomas Misa's *Nation of Steel: The Making of Modern America, 1865–1925* and Robert Gordon's *American Iron, 1607–1900* are thoroughly researched texts. Although they are written for academic audiences, both of these books are enjoyable to read. One of the most important books on the impact of the steel rails is *The Railroad Journey: The Industrialization of Time and Space in the Nineteenth Century* written by Wolfgang Schivelbusch. This small, erudite book should be required reading for any engineering ethics or sociology classroom. There is a thicker and more general book on the impact of the rails called *The Age of the Railway* by Harold Perkin, which has a British focus but is comprehensive in its coverage. A discussion of the annihilation of space is the focus of Barney Warf's appropriately titled *Time-Space Compression: Historical Geographies*.

The Commercialization of Christmas. There are many books about the railroads, but the discussion on the role of railroads on the commercialization of Christmas is limited. Historian Penne L. Restad unpacks the connection in her article "Christmas in America: A History"; and the book *Christmas: A Candid History* written by Bruce D.

Forbes also considers this link. Scholars might find it interesting to examine newspaper clippings, which show the evolution of Christmas from a minor holiday to the modern one we experience today.

Chapter 3: Convey

The Battle of New Orleans. Any serious reader of Andrew Jackson should pick up a book written by his primary biographer, Robert V. Remini, specifically the one entitled *The Battle of New Orleans: Andrew Jackson and America's First Military Victory*. But the reader should take care not to be too mired in the point of view of this one prolific author. There are other books that also provide extensive accounts, augmenting Remini's point of view, which include *A British Eyewitness at the Battle of New Orleans* edited by Gene A. Smith, and *Glorious Victory: Andrew Jackson and the Battle of New Orleans* by Donald R. Hickey.

In addition to these titles, the war has been discussed in an older book by Francis F. Beirne, called *The War of 1812,* which is written in an accessible style. For the British point of view, the book *The British at the Gates: The New Orleans Campaign in the War of 1812* by Robin Reilly is a must-have and is filled with facts not typically found in American depictions. For those readers seeking a more visual treatment of the war, *The Rockets' Red Glare: An Illustrated History of the War of 1812* by Donald R. Hickey and Connie D. Clark is a wonderful illustrated volume that shows the key players involved, as well as elaborate maps. This title is very well done and provides a great account of what transpired. Also, readers will find that the chapter on the battle of New Orleans in *What Hath God Wrought: The Transformation of America, 1815–1848* by Daniel Howe possesses fresh details as well. Classrooms might also enjoy documentaries on this battle (particularly ones produced by the History Channel and PBS).

For the serious scholar who wishes to read Jackson's papers, the Library of Congress has over 20,000 items. The Hermitage, Andrew Jackson's home, also serves as a repository for his archives and offers digital versions of his papers as well. Newspaper accounts of the battle are also highly informative; in particular *The Niles' Weekly Register* (a Baltimore newspaper) provides a ringside seat to what occurred on that Louisiana plantation. Lastly, a history buff would enjoy a visit to the Chalmette Battlefield in New Orleans, particularly on the anniversary of the battle in early January.

Samuel F. B. Morse. Samuel Finley Breese Morse was a much-revered inventor and his story was a tale known well by schoolchildren over a century ago. As such, many of the books written about him are of an older vintage. Only recently has there been a contemporary biography about him, authored by Kenneth Silverman and named *Lightning Man: The Accursed Life of Samuel F. B. Morse*. This is a thick tome, crafted by a consummate biographer, and chock full of details. A serious reader would do well to pick up a copy of this book.

A modern telling of the Morse tale that is shorter than 300 pages does not exist. One will have to go back to older books such as *Samuel Finley Breese Morse* by John Trowbridge published in 1901. At 134 pages, it is a brisk read that condenses Morse's flaws into one sentence. This book provides a nice timeline; however, there is an error in the date that the telegraph was erected (which was 1844). A longer read about Morse is the Pulitzer Prize–winning *American Leonardo: A Life of Samuel F. B. Morse* by Carleton Mabee, which is written in a journalistic style that would still be appreciated by modern readers today. Unfortunately, there are some errors in this book as well. But to its credit, the book has an entire chapter called "Native Americans," which talks about Morse's political, nativist side. A short, 215-page biography called *Samuel F.B. Morse and the American Democratic Art* by Oliver Waterman Larkin is a quick and enjoyable read, but quite hard to acquire. Another account is *Wiring a Continent: The History of the Telegraph Industry in the United States* by Robert Luther Thompson. This book is desirable for those wanting a depiction of the telegraph without the fat of character and story.

The book *The Life of Samuel F.B. Morse* by Samuel Irenaeus Prime is the reference that many other books rely on and cite. Samuel Prime was chosen by the family to write this biography, so he had access to materials that are not available to many others. Prime's is a good text that is filled not only with correspondences that might otherwise be hard to acquire, but drafts of depositions as well as smart commentary. This book is quite technical both from the scientific point of view and from a legal one. There is very little description of Morse's work as an artist in its pages. Nevertheless, this is a must-have text for those wishing to get a full picture of the development of the telegraph. To explore Morse the artist, it is suggested that readers review the large-format book entitled *Samuel F. B. Morse* by William Kloss, which celebrates Morse's art, cataloging his paintings in full color as well as critiquing his skill.

Morse's letters reside in a thick two-volume set called *Samuel F. B. Morse: His Letters and Journals in Two Volumes,* compiled by Samuel Morse's youngest son, Edward Lind Morse. Each of Samuel Morse's two lives—as an artist and as an inventor—lasted for about forty-one years; volume 1 covers Morse's early life as a school boy, as a young artist, and as a newly married man, while volume 2 is about Morse as the inventor, his time on the *Sully*, his creation of the telegraph, and the development of the first working line. This is the volume to acquire if the full set is not available.

Overall, for the discussion of the development of the telegraph, a combination of the works of Silverman, Mabee, Lind Morse, and Prime will serve any serious researcher looking for diverse takes on the story, as well as provide an array of information. As with any book one reads, it is important to double-check the dates and details with earlier sources, since there are errors that migrate from one generation of books to the next.

A student or scholar of Morse would be pleased to find that Morse was a prolific letter writer and that many of his letters are freely available on the Library of Congress website. The Samuel Finley Breese Morse Papers (MSS33670) will make those who love lots of materials weep. The archives at Yale University are much smaller, but there are key letters written by Morse, particularly as he attempted to get a British patent. The most important collection on Morse at Yale is the Yale College Alumni file (RU 830, Box 2), which has articles written about Morse closer to the time of his work. It should be noted that the New York Public Library also has some of Morse's letters, which are available online.

For modern mentions of the tale of Samuel F. B. Morse, some excellent books weave Morse's activities into the larger fabric of the development of telecommunications. One book that stands out is the *Victorian Internet: The Remarkable Story of the Telegraph and the Nineteenth Century's On-line Pioneers* by Tom Standage. This book is a champion not only for its fantastic title, but for its refreshing and engaging telling of the birth of the telegraph, telephone, and wireless. Another title that was less popular but equally as engaging is *The Electric Universe: How Electricity Switched on the Modern World* by David Bodanis. Again, this title brings characters to life in an admirable way, making dry details exciting and explaining our knowledge of electricity and how we have employed it to create our modern world. Both of these titles will serve anyone generally interested in the development of telecommunications (and the characters that made it possible). In addition to those great examples of exposition, there is the older *The Story of Communications from Beacon Light to Telstar* by Senator John Pastore, which is a short survey providing a quick read.

James A. Garfield. Garfield's presidency was short, so little has been written about him. Fortunately, a biography of James Garfield by Candice Millard called *Destiny of the Republic: A Tale of Madness,*

Medicine and the Murder of a President was recently published. This well-researched and well-written book was the foundation for a PBS documentary. Overall, Garfield never really got his due. Of the few books about him, most of them, like his presidency, are brief, but together allow a reader to get a fuller picture of the man. The thin book entitled *James A. Garfield* by Edwin P. Hoyt published decades ago, adds lots of color about the former president, ending with the assassination. This book describes key stages in Garfield's early life, with more details than longer books. Another short book is *James A. Garfield* by Ira Rutkow and Arthur M. Schlesinger, Jr. This is a book that any fan of medicine should obtain. Rutkow, a professor of surgery, fills in the blanks about what killed Garfield and the state of medicine in 1881 better than any other texts on the market.

The collections of Garfield's papers are located in the Library of Congress and at the Hiram College in Ohio. However, some of them are bound in a two volume-set entitled *The Life and Letters of James Abram Garfield* by Theodore Clarke Smith. The chapter in the second volume, entitled "The Tragedy," gives a thorough account of the shooting of the president in less than thirty pages. Another book, entitled *The Murder of James A. Garfield: The President's Last Days and the Trial and Execution of His Assassin* by James C. Clark, is a bit harder to obtain, but its prologue is downloadable on the National Archives website. The *New York Times* article "A Great Nation in Grief," printed on July 3, 1881, carries a wealth of information and eyewitness accounts. Many of the books listed above used this article as a source.

Fans of history would be keen to know that the bullet shot by Guiteau, a vertebra from Garfield's spine, and a piece of Charles Guiteau's brain are located at the National Museum of Health and Medicine. Reports of Garfield's autopsy can be found in the *Complete Medical Record of President Garfield's Case, Containing All of the Official Bulletins,* published by C. A. Wimer, 1881.

The Telegraph. Just as the telegraph compressed language into a code, a little book on the history of the telegraph concentrates this history in a similar fashion. Each of the sentences of Lewis Coe's *The Telegraph: A History of Morse's Invention and Its Predecessors in the United States* could be lengthened to a paragraph, but this rich book gives the reader a broad sense of the impact of Morse's invention. For more information about the telegraph, *The Story of Telecommunications* by George P. Oslin provides plenty of details. This thick book offers something for everyone—from great pictures to an encyclopedic format. While the exposition can be a bit choppy, this resource is unparalleled in its scope. There is also *The Electric Telegraph: A Social and Economic History* by Jeffrey L. Kieve, which has a British focus. The more recent book *The Telegraph in America, 1832–1920* by David Hochfelder is a well-researched and thorough book, targeted for academic audiences. Readers wanting to know about the linkage between the telegraph, language, and journalism, however, should read chapter 3; they will come away quite informed.

For information on the evolution of language and the role of technology, *Alphabet to Email* by Naomi S. Baron is a must-read. The discussion on the telegraph is short, but the writing is broad, inviting, and refreshing. Baron's book can be supplemented by *Patriotic Gore* by Edmund Wilson. This is a treasure, for it takes a snapshot of language around the time of the birth of the telegraph. Its premise is that the mechanical age, along with the Civil War, was the impetus for a "chastening of language," whereby one factor was the telegraph. The impact of the telegraph can also be explored by examining how it changed the way news was consumed. *News over the Wires: The Telegraph and the Flow of Public Information in America* by Meneham Blondheim is an erudite examination of the history of the news service, from the birth of the telegraph to the creation of the Associated Press. In its pages are great details about how news was transported before and after the advent of the telegraph.

As for today's modes of online communication, a few titles worth mentioning have bubbled up. For the impact of our devices and social media, Sherry Turkle's *Reclaiming Conversation: The Power of Talk in a Digital Age* pulls the fire alarm alerting society to the insidious nature of these forms of communication. Despite this stance, her work is optimistic and posits that modern society's loneliness, caused by social media, can be mitigated by making space for face-to-face conversations. The impact of instant communication was also described in the early work of Lewis Mumford, *Technics and Civilization*. In its pages, Mumford made sound predictions, particularly noting that empathy and sympathy will be hard to transmit and receive—both of which are modern concerns for educators, parents, and scholars.

Chapter 4: Capture

Eadweard Muybridge. For accounts about Eadweard Muybridge, a reader should look no further than Rebecca Solnit's literary masterpiece, *River of Shadows: Eadweard Muybridge and the Technological Wild West*. This is a well-researched, transcendent, and beautiful-to-read narrative that describes history by using Muybridge as the connective tissue. Solnit's book is exemplary, but it is not the only admirable offering in the telling of Muybridge's story. For another wonderful dose of writing, there is also *The Inventor and the Tycoon* by Edward Ball, a sweeping tome that covers the Stanford–Muybridge connection and all of the characters and events in Muybridge's life. Some readers do not require great storytelling, however. For a scholar seeking hard-to-find facts, *Eadweard Muybridge: The Stanford Years (1872–1882)* by Arthur Mayer is a desirable choice. Also, the recent book by Marta Braun entitled *Eadweard Muybridge* published in the UK provides a fresh perspective, and although it is thin it is thorough.

Many readers enjoy a well-told murder story. The book *A Million and One Nights: A History of the Motion Picture,* written by Terry Ramsaye, takes the position that Muybridge made no contribution to moving pictures; however, this author spells out the entire Muybridge homicide saga with exquisite and engaging detail, in the style of a murder mystery. For a reader who wants to absorb a tale well told, this is it. For murder trial accounts, the papers from the Napa Historical Society will grip a reader, too.

Other useful texts are Robert Bartlett Haas's *Muybridge: Man in Motion* and Brian Clegg's *The Man Who Stopped Time.* Additionally, the California Digital Newspaper Collection (https://cdnc. ucr.edu/) has many Golden State newspapers online. Details about Muybridge's earliest attempts to capture a horse in motion as well as the murder case can be found there. There is also a comprehensive website maintained by Stephen Herbert called The Compleat Eadweard Muybridge (http://www.stephenherbert.co.uk/muyb COMPLEAT.htm). Lastly, Muybridge wrote several books. One that is commonly found in most libraries is *Animals in Motion.* In it, one will find details about his outdoor camera studio as well as a large catalog of photographs taken with his unique photographic setup.

Hannibal Goodwin. For such an important inventor, very little has been written about the good Reverend Goodwin. One article from 2001 by Barbara Moran, "The Preacher Who Beat Eastman Kodak," depicts the story of Goodwin and his battle against Eastman, and was published in the now-defunct magazine called *Invention and Technology.* Goodwin was also the focus of a short (13-page) monograph by George Helmke called *Hannibal Goodwin and the Invention of a Base for Rollfilm.* This title is rare in most libraries, but it is invaluable. Copies can be acquired from the Fleetwood Museum of Art & Photographica of the Borough of North Plainfield, New Jersey, which published it. One can find details about Hannibal Goodwin in Robert Taft's *Photography and the American Scene*. Additional materials can also be found in the *Cyclopedia of New Jersey*, and also in the lengthy book *George Eastman: A Biography* by Elizabeth Brayer, but this account is very sympathetic to Eastman. A good summary of the legal battle between the preacher and the mogul can be found in H. W. Schütt's article "David and Goliath: The Patent Infringement Case of Goodwin v. Eastman."

A serious scholar of Goodwin would do well to visit the Charles F. Cummings New Jersey Information Center in the Newark Public Library, as well as peruse the Charles Pell Papers at the New Jersey Historical Society, also in Newark. Newspaper clippings in the former and correspondences in the latter will help to fill in any blanks, particularly Goodwin's formal statement in the Pell Papers. The George Eastman Archives has scores of letters that pertain to Goodwin (although none are from Goodwin). The legal documents, which are thicker than bibles and heavier than a table, are located there too. For information about the material that Goodwin used, there is a fun book called *Cellulose, The Chemical That Grows* by William Haynes, which gives an understandable description of the history and use of this once popular chemical compound. Robert D. Friedel has also written a short book called *Pioneer Plastic: The Making and Selling of Celluloid*, which describes the history of a material that is now mostly forgotten.

Frederick Douglass. Frederick Douglass was a big fan of photography and even now a picture of him occasionally pops up in an old scrapbook found in an attic (the Rochester Library has one such photograph). Frederick Douglass made several speeches effervescing with his infatuation with photography. The book *Picturing Frederick Douglass* by authors John Stauffer, Zoe Trodd, and Celeste-Marie Bernier contains a transcription of three of his key speeches where he describes his affinity for this art form, as well as over 150 images of Douglass himself. While Douglass's speeches "Lecture on Pictures,"

"The Age of Pictures," "Life Pictures," and "Pictures and Progress" can be found written in Douglass's hand on the Library of Congress website, his notes may be difficult to read for many. As such, the transcriptions in *Picturing Frederick Douglass* make this book a tremendous resource. An approachable article on Frederick Douglass's use of photography can be found in Henry Louis Gates's essay in this volume. Another resource that is useful in the discussion of Frederick Douglass's use of photography is the well-researched book *The Camera and the Press* by Marcy J. Dinius.

There has been a resurgence in the study of Frederick Douglass's abolitionist speeches not only in America, but also in Great Britain, where Douglass spent a few years away to sway public opinion about slavery. By using the British press, which was often covered by the American press, Douglass was able to distribute his message of abolition circuitously. The work of Dr. Hannah-Rose Murray speaks to this. At the time of this printing, a website held her research (frederickdouglassinbritain.com), with a fantastic map of all the places that Douglass visited across the pond.

Shirley Cards. Specifics about Shirley Cards can be found in Lorna Roth's seminal work "Looking at Shirley, the Ultimate Norm," a paper published in the *Canadian Journal of Communication* in 2009. What makes Roth's work so groundbreaking is that she garnered several interviews and notes from former Kodak executives and employees. This paper should be required reading for all scholars and educators on technology. Photographers Adam Broomberg and Oliver Chanarin have brought this work to public attention. A report in *The Guardian* about their exhibit, called "'Racism' of Early Colour Photography Explored in Art Exhibition" posed the question "Can the camera be racist?" This torch has been carried by the book *Technically Wrong: Sexist Apps, Biased Algorithms, and Other Threats of Toxic Tech* by Sara Wachter-Boettcher. What we know from these works and from Professor Roth's insight is that there are biases inherent in the technologies we hold dear.

Polaroid. Very little has been written about Caroline Hunter, Ken Williams, and the Polaroid Revolutionary Workers Movement (PRWM), except for entries in books about Polaroid that describe both Caroline and Ken as fanatical employees. These books include Mark Olshaker's *The Instant Image: Edwin Land and the Polaroid Experience* and *Land's Polaroid: A Company and the Man Who Invented It* by Peter C. Wensberg. Wensberg was an executive at Polaroid and was present when the PRWM events transpired. But Wensberg was also a company man and wrote his book in that light. Interestingly, books on Polaroid written in the twenty-first century do not mention this significant bit of Polaroid history, as some authors choose to explore the joy of instant photography without its well-documented social impact. These new titles are cases of revisionism or lazy journalism or both.

Some academic papers tell the story of the PRWM, such as Eric J. Morgan's "The World Is Watching: Polaroid and South Africa." Also, there are documentaries that make mention of this episode in history, such as "Have You Heard from Johannesburg?" with a cameo by Caroline Hunter. Caroline Hunter was also interviewed more recently on *Democracy Now!* in 2013. For footage of Caroline Hunter in the 1970s, she was a guest on *Say Brother* on WGBH in Boston.

To acquire more details about the PRWM, the *Harvard Crimson* depicts what happened during the time of the events. There are also archival materials, located on the Michigan State University website, called the African Activist Archive Project, available at www.africanactivist.msu.edu. The PRWM archives can be found within the New York Public Library's Schomburg Collection for Research in Black Culture in Harlem and are a tremendous resource for this mostly undocumented chapter in history. The Polaroid Corporation archives, found at the Harvard Business School, would also be invaluable to a scholar researching this subject.

Chapter 5: See

William Wallace. The inventor William Wallace is often left as a footnote in most history books about Edison. Even modern books continue this trend. Fortunately, engineer and Edison fan William Hammer, who was a great chronicler of Edison's inventions, wrote three pieces about William Wallace for the *Electrical Engineer*. These pieces were published in 1898 and can be readily found. There is also a mention of Wallace in volume 3 of *The Papers of Thomas A. Edison*, published by the Johns Hopkins University Press. And there are a few newspaper clippings on Wallace marking his death, and local papers in Connecticut occasionally make mention of him. Alas, there is very little written about the man who catalyzed Edison's electric lights. Luckily, the most exhaustive collection on William Wallace can be easily acquired from the Derby Historical Society in Connecticut, which has many of these clippings and William Hammer's articles. They have a very small vertical file on Wallace, as well as pictures of him. One of Wallace's arc lamps still resides in Ansonia, but it is part of a private collection. The Smithsonian has one of his arc lamps, too, as well as Wallace's telemachon.

Edison's Electric Lights. The development of the light bulb has been thoroughly documented in one of the best books on the topic, *Edison's Electric Lights: The Art of Invention* by Robert Friedel and Paul Israel. The earliest edition of this book is the best volume to obtain, for it includes more pictures than the more recent version. Accounts of the development of the light bulb can be found there, as well as in a number of biographies about Edison. Some of these include Neil Baldwin's *Edison: Inventing the Century*, George Sands Bryan's

Edison: The Man and His Work, Robert E. Conot's *A Streak of Luck*, Frank Dyer and Thomas Martin's *Edison: His Life and Inventions*, and Matthew Josephson's *Edison: A Biography*. This last book outshines the others in the discussion of the light bulb. It is an older tome that discusses Edison's thinking for building an electrical system based on lights in series or in parallel. For a shorter portrayal of the development of the electric lights, there is also *Edison: The Man Who Made the Future* by Ronald Clark, which has a short chapter on the birth of the light bulb. Many of Edison's papers are on the Rutgers University website, which is a tremendous resource as well. Archival materials about the development of the electric light can be found in the William Hammer Collection in the Smithsonian Institute under the topic of Edisonia. Hammer did our nation a great service by saving every report and newspaper article about Edison.

For a discussion of the life of the electric light after Edison, the book *Lamps for a Brighter America: A History of the General Electric Lamp Business* by Paul W. Keating is a good start. For the history of lighting, there is also Brian Bowers's *A History of Electric Lights and Power*. This book stands out in giving an account in how artificial illumination came to be.

For readers who are serious fans of Edison and want to be in the space where the invention occurred, a visit to Menlo Park as it once looked is possible. The building is no longer in New Jersey, however, but located in the Henry Ford Museum in Dearborn, Michigan. Henry Ford admired Edison so much that he brought over the whole building, including some of the soil. Inside Menlo Park, the furnaces on the first floor were used to make carbon filaments. On the second floor, the vacuum pump used to evacuate glass bulbs sits in this room with walls full of jars. Menlo Park is worth a visit and a must see for any scholar of Edison.

Light and Society. The role of artificial illumination in society has been a topic written about numerous times and in numerous ways. The most seminal—and eye-opening—work on the impact of artificial lights on our culture can be found in Wolfgang Schivelbusch's *Disenchanted Night: The Industrialization of Light in the Nineteenth Century.* This book is a must read, rich with information and thought-provoking ideas. Other contributions to the canon of artificial light ranges from the beautiful prose of Jane Brox's *Brilliant: The Evolution of Artificial Light* to the technical yet readable *City Lights: Illuminating the American Night* by John A. Jakle. David E. Nye's *Electrifying America: Social Meanings of a New Technology* unpacks the social impact of light and electricity, and his examination is considered a model work.

As for the topic of light pollution, this subject has been well documented within the scientific literature. Some of these treatises have crossed the academic Rubicon over to the general public. One such book, *Ecological Consequences of Artificial Night Lighting*, is edited by Catherine Rich and Travis Longcore. Some of the information about the impact of artificial lights on wildlife—and on humans— has made its way from books like this into other books, articles, and the news. The most readable and enjoyable book on the loss of the night is Paul Bogard's *The End of Night: Searching for Natural Darkness in an Age of Artificial Light*, for it is both well researched and clearly written. In some instances, it is even lyrical, as it shares insights and warnings about the loss of an old friend of humanity—the dark. For readers who just want the facts, the short book by the International Dark-Sky Association called *Fighting Light Pollution* discusses the consequences of light pollution, as well as what each of us can do to reduce it.

Chapter 6: Share

The Golden Record. A solid account about the creation of the Golden Record resides within the pages of *Murmurs of Earth: The*

Voyager Interstellar Record, a compilation of essays by Carl Sagan, F. D. Drake, Ann Druyan, Timothy Ferris, Jon Lomberg, and Linda Salzman Sagan. How the record came to be and what is on the record are all provided there. While the work to make the Golden Record all happened in the late 1970s, updated articles have been written to coincide with the fortieth anniversary of the launch of the Voyager spacecrafts. These offerings include a chapter in the book *Interstellar Age: Inside the Forty Year Voyager Mission* by Jim Bell, the article "How the Voyager Golden Record Was Made" by Timothy Ferris, written for the *New Yorker,* and the insert accompanying the reissued compact discs produced by Osma Records. Stories about the creation of the record are also in biographies about Carl Sagan, including Keay Davidson's *Carl Sagan: A Life* and William Poundstone's *Carl Sagan: A Life in the Cosmos.* Interestingly, the Golden Record has been the focus of academic dissertations, such as William Macauley's thesis (in the UK); children's books, such as *Star Stuff;* and documentaries, such as PBS's *The Farthest,* which is worth watching. While the Golden Record is older than most Americans, it still continues to captivate.

On the NASA Jet Propulsion Laboratory website, a reader can find pictures of how the record was made. For the curious and committed, archival materials about the Golden Record can be found online in the Seth MacFarlane Collection of the Carl Sagan and Ann Druyan Archive from the Library of Congress. But this is only a small sampling of what is available. The majority of the materials are hardcopies located in the Library of Congress and require a visit to Washington, DC. Unfortunately, this collection does not contain an actual Golden Record (only a few were made), but the drawings and letters will illuminate how thrilling—and stressful—it was to put this interstellar compilation together.

Alan Lomax. Alan Lomax was an American treasure, for he collected the songs that meant something to all parts of the world. Lomax had a long and wide-ranging career, but all of the materials that pertain to Lomax's involvement in the Golden Record are available online on the Library of Congress website. The most readable entry about the songs that Lomax selected can be found in the blog entitled "Alan Lomax and the Voyager Golden Records" written by Bertram Lyons. From this article, which was posted on the Library of Congress website in 2014, one can see the fifteen out of twenty-seven songs Lomax selected for the Golden Record. Other materials corroborating that list can be found in the Sagan–Druyan papers.

For information about Alan Lomax and his work, a few books focus specifically on him. There is the biography written by John Szwed called *Alan Lomax: The Man Who Recorded the World*. There is also a multiauthored book called *The Southern Journey of Alan Lomax*. To understand how Lomax ticked, one has to read his beloved work called *Cantometrics*. Here, Lomax presented a graphic representation of each song—similar to an EKG—based on a thirty-seven-point classification of musical styles (of tempo and rhythm, phrasing and polyphonics) that he devised as he attempted to make his work more scientific with this systematized taxonomy. Just as Sagan read graphs on stars, Lomax, too, had graphs for music. But Lomax's work never got the attention he believed it deserved. Be that as it may, Lomax wrote many articles and created a vast collection of music. Many of these items can be found in the Alan Lomax Archives in the Library of Congress.

Edison's Phonograph. Many books give an account of the origin story of the phonograph, sometimes providing overlapping details. A few of them include Neil Baldwin's *Edison: Inventing the Century*, George Bryan's *Edison: The Man and His Work*, Robert E. Conot's *A Streak of Luck*, Frank Dyer and Thomas Martin's *Edison: His Life and Inventions*, and Matthew Josephson's *Edison: A Biography*. There is also a small and readable text called *Edison: The Man Who Made the Future* by Ronald W. Clark that has an entire chapter on the phonograph. Among these books, Conot's contemporary account stands out from the others, benefiting as it does from the efforts of earlier books as well as the author's own research.

Sadly, the phonograph was eclipsed by the light bulb. Had the phonograph been created by a lesser inventor, it would have been written about more. There are a few books to fill this void, such as *The Fabulous Phonograph* by Roland Gelatt, *From Tin Foil to Stereo: Evolution of the Phonograph* by Oliver Read and Walter L. Welch, and *The Talking Machine: An Illustrated Compendium, 1877-1929* by Tim Fabrizio and George F. Paul. Together, these works can help paint a more complete history of the phonograph and its impact.

To get the whole picture about the development of the phonograph, readers can peruse Edison's laboratory notebook entries without having to visit New Jersey. The papers that pertain to the creation of the phonograph are included in volume 3 of *The Papers of Thomas A. Edison*, published by the Johns Hopkins University Press, which is more extensive than what is provided online by the Thomas Edison Papers at Rutgers University (http://edison.rutgers.edu/). This particular volume includes his work from April 1876 to December 1877. Notebook entries contain a hodgepodge of ideas and drawings, but one can get a sense of the dates as well as his other activities. In the appendix to the third volume is an account of the development of the phonograph by Charles Batchelor, Edison's assistant, but it was written nearly thirty years after the invention. As such, Batchelor conflates activities that happened over the course of months into just a few days. An account that is as close as we will get to hearing Edison's telling of the story is George Parsons Lathrop's "Talks with Edison," published in *Harper's Weekly* in 1889. Again, this account was published twelve years after Edison's invention, but it includes quotes from the wizard himself.

Edison had great plans for his favorite invention, which are stated in a piece in the *North American Review* called "The Phonograph and Its Future," published a year after the phonograph's invention, in 1878. While he was a marvelous inventor, he was not a good futurist, for he did not see the full potential of the phonograph for music. Despite this, the article is fun to read, as most of the things that he did predict came true by the end of the twentieth century. Another resource is the phonograph patent itself (No. 200,521), as well as the 1877 article in *Scientific American*, which not only broke the story, but was a key part of the phonograph's history as well.

Recording Technology's History and Impact. James Gleick's *The Information: A History, A Theory, A Flood* chronicles how data was stored, beginning from the days of marks in clay to today's computers. It is a thoroughly researched book that will leave no reader wanting. The story of the development of the science behind data has been overlooked for a long time and it now has one of the best writers bringing it to light. What remains missing in the canon of information is a discussion of the role of magnets in data storage and on society in general. At the time of this writing, there is the book *Driving Force* by James D. Livingston, and there are many articles written by scientists for technical audiences, but magnets by themselves have not been profiled by such confident hands as Gleick's. As such, magnets still remain a mystery to most and their use in our culture is taken for granted. Magnets have undergirded society from compasses, to hard disks, to medical research. It is hoped that a writer will take up the worthy cause of bringing magnets into the limelight they deserve.

In addition to magnets being missing from discussions about information storage, so too is the tinfoil from Edison's phonograph missing from discussions about recording materials. Many specialized books on recording media overlook Edison's tinfoil and begin their discussions with Valdemar Poulsen's wires impregnated with iron filings. Magnetic media has certainly held the lion's share of the world's recordings, but before this medium existed, data, starting with sound, was recorded on tinfoil wrapped around a cylinder.

This fact gets lost as one author follows another author and so on. But a thoughtful examination should include Edison's work. One of the premiere universities in the study of magnetic recording, UC-San Diego, recognized Edison's efforts on their website of Recording Technology History in notes written by Steven Schoenherr in 2005.

Overall, there is a need to include the recording of sound in the canon of information storage. The impact of the ability to record sound has been discussed in a few books. One very accessible title is *Infoculture*, which comes from the Smithsonian and was written by Steven Lubar. Another fine tome is *America on Record: A History of Recorded Sound* by Andre Millard, which discusses the history of storing sound as well as the impact of these means of storage on American life. Details on the role of magnetic media can be found in James Livingston's "100 Years of Magnetic Memories," which explores how the ability to record sound not only influenced music, but led to the impeachment of President Nixon. This is a breezy, short article that provides a timeline of key events. This article gives a great overview, but to gain a deeper understanding of the science behind magnets will require that a reader peruse thicker books, such as *Magnetic Sound Recording: Theory and Practice of Recording and Reproduction* by D. A. Snel, or a highly technical tome, such as B. D. Cullity's *Introduction to Magnetic Materials*.

Data and Privacy. The social, legal, and ethical issues of computers, the internet, and data have been best spelled out in a textbook written by Sara Baase called *A Gift of Fire* (a twist on the story of Prometheus). A keen reader or scholar will find its clear exposition, legal cases, and references to be highly useful. For books that are targeted for general audiences, there is the cleverly titled *Data and Goliath: The Hidden Battles to Collect Your Data and Control Your World* by Bruce Schneier and *Big Data: A Revolution That Will Transform How We Live, Work, and Think* by Viktor Mayer-Schönberger and Kenneth Cukier. The venerable Very Short Introduction series has a volume titled *Privacy* written by Raymond Wacks, which is a good read and is published by Oxford University Press.

Chapter 7: Discover

Penicillin. The story of penicillin begins with Alexander Fleming's discovery of a mold in a petri dish. But his observation that this mold killed germs was just the start. In order for penicillin to be a useful antibiotic for people, the mold had to be grown in massive amounts. That work was done by Oxford scientists Howard Florey, Ernst Chain, and Norman Heatley, and biographies of these scientists as well as Fleming provide the whole story of penicillin.

Two contemporary books will benefit any reader. *The Mold in Dr. Florey's Coat: The Story of the Penicillin Miracle* by Eric Lax is a well-researched account and a fine example of storytelling. In writing his book, Lax gained access to some of Heatley's rare personal materials as well as those of other scientists, making for a richer story. Another authoritative book is Kevin Brown's *Penicillin Man: Alexander Fleming and the Antibiotic Revolution*. Brown is a historian and also the curator at the Alexander Fleming Museum in London. As such, Brown possesses a deep understanding of Fleming's life and work, and has painstakingly pieced together rare materials. Brown's book, along with Lax's title, are certainly worth acquiring. Other biographies pertaining to penicillin include older texts such as *Alexander Fleming: The Man and the Myth* by Gwyn Macfarlane and *Howard Florey: The Making of a Great Scientist* also by Gwyn Macfarlane. Macfarlane is a fine writer, but, with one author writing about two very important figures, the story can become uneven; a thoughtful reader must take care to acquire other books. One such book to balance Macfarlane's accounts is Lennard Bickel's *Rise Up to Life: A Biography of Howard Walter Florey Who Gave Penicillin to the World*.

A general discussion of the development of penicillin can be found in a number of books that aren't biographies. The short book *Yellow Magic: The Story of Penicillin* by John Drury Ratcliff was written around the time that penicillin was discovered and gives the reader a sense of how the world viewed this achievement. Also, John C. Sheehan's *The Enchanted Ring: The Untold Story of Penicillin* focuses on Sheehan's work in the latter part of penicillin's development. This book also discusses the Cocoanut Grove fire in Boston in 1942, where penicillin became famous in the United States for saving many burn victims. In addition to Sheenan's book, Robert Hare's *The Birth of Penicillin* debunks the myth that the spore of mold came from the window. Hare claims it actually came from the laboratory on the first floor. For those who would prefer to watch the story of penicillin unfold instead of reading about it, the movie *Penicillin: The Magic Bullet* (2006) brings Florey's story to the big (and little) screen.

Penicillin saved millions of lives and the Nobel Prize of 1945 went to Alexander Fleming, Ernst Chain, and Howard Florey. Norman Heatley, an unsung hero, was left out. The clever Heatley was the key to the manufacture of penicillin. When World War II prevented the use of real scientific equipment to make large quantities of penicillin, Heatley—a master improviser—used bookcases and bedpans to manufacture it in the amounts required. Unfortunately, Heatley never really got the recognition he deserved. Some authors have made solid efforts to correct this, however. A short, self-published, and highly readable book called *Penicillin and the Legacy of Norman Heatley* by David Cranston and Eric Sidebottom discusses Heatley's work in coaxing penicillin out from the mold. Heatley also wrote about his efforts in *Penicillin and Luck,* and his laboratory notebooks and journals housed in the Wellcome Trust are enjoyable to read. Notwithstanding, he deserves much more recognition for his contributions.

Scholars who desire more information than that contained in these books will be pleased to know that much of the original materials is accessible in archives. Fleming's papers are located in the British Library and Ernst Chain's papers are in the Wellcome Trust Library along with Heatley's. Some of Florey's papers are located in the Royal Society archives and some are at Yale University. At Yale, papers from the John F. Fulton collection are useful, too, for Fulton was Florey's colleague and friend. Lastly, details about the first US citizen to receive penicillin are located in the Yale Medical Library.

Glass. A few nontechnical books about glass have been published over the last few decades. A recent title that is targeted for general audiences is called *Glass* and authored by William S. Ellis, who tells the curious story of this material from its ancient beginnings to its modern use in optical fibers. The beautiful book *Glass: 5,000 Years* by Hugh Tait provides an illustrated history of glass and offers many colorful specimens of ancient glasswork. Both lay readers and serious glassblowing aficionados would benefit greatly from this book. What is rare in the literature is a step-by-step guide of how to blow glass. Those interested in glassblowing will be pleased to see such a guide at the end of Tait's book. For those readers who might not want an examination of the aesthetics of glass, but are wanting a bit more technical information about glass, the exhaustive book *Glass: The Miracle Maker* by C. J. Phillips, an older text, remains a perennial classic. Far more technical books than this title exist, but this book discusses the history of glass and also its technological applications. A serious student of glass would welcome this old book on their shelf. For readers interested in the role of glass in the sciences, one useful scientific paper on this topic is "Glass: The Eye of Science" written by Marvin Bolt.

Pyrex. Little has been written on Otto Schott, and even less in English. A basic biography can be found on the Schott Glass website.

One example is the article "From a Glass Laboratory to a Technology Company" published in the 2009 edition of *Schott Solutions*. In addition to these articles, a few scientific papers discuss the life of Otto Schott. A key biographical summary is "Otto Schott and His Work" by W. E. S. Turner, which was written in 1932. Professor Turner was able to acquire materials from the Schott family, as well as have his summary reviewed by Schott's son. Another seminal paper is "Otto Schott and the Invention of Borosilicate Glass" by Jurgen Steiner, an employee of Schott Glass. This is one of the most exhaustive descriptions of Schott's work and contains forty-five references (most in German).

Information about the development of Pyrex in the United States can be found in various scientific papers, scholarly works, and a popular book. On the science side, there is the original Pyrex paper, which is entitled "The Development of Low Expansion Glasses" written by E. C. Sullivan. One book that describes the historical development of Pyrex is *Corning and the Craft of Innovation* by Margaret B. W. Graham and Alec T. Shuldiner. This book was supported by Corning and should be read critically from that point of view. There is also *The Generations of Corning* by Davis Dyer and Daniel Gross, which provides the most comprehensive historical account of the origin of Pyrex. Lastly, Regina Blaszczyk's *Imaging Consumers: Design and Innovation from Wedgewood to Corning* mentions Pyrex's development in a limited way. Overall, the story of Pyrex is still waiting for a full scholarly examination.

For a brief write-up about Pyrex for general audiences, there is a fine short paper called "The Origin of Pyrex" by William B. Jensen. This is a highly readable, but very concise, account. An article from the 1949 *Gaffer Magazine* called "The Battery Jar that Built a Business" is a good resource that can be requested from the Corning Incorporated archives. Additionally, the Corning Museum of Glass has several short historical summaries and bibliographies written about the development of Pyrex on their website. In

2015, this museum also had an exhibit celebrating Pyrex's 100th anniversary.

For details about Bessie Littleton, one of the best accounts of her life—and personality—can be found in a self-published book by her son Joseph C. Littleton called *Recollections of Mom*. This book can be acquired from the Rakow Research Library (of the Corning Museum of Glass). Also, the Smithsonian Archives of American Art has an oral history of the famous glass artist, Harvey K. Littleton, Bessie and J. T.'s son, which also has some elements about the origin story of Pyrex. Interestingly, there is a mention of Bessie Littleton in Mary Roach's *Bonk: The Curious Coupling of Science and Sex*.

An examination of the Trading with the Enemy Act, which included many of the technologies that the United States uses, from aspirin to borosilicate glass, still needs to be undertaken. Most textbooks never mention this, and most of the discussions are tied up in scholarly work by historians of economics. The *Scientific American* article "Trading with the Enemy Act" from 1917 discusses the booty of war. State archives, such as those in New York, have long lists of products that became available to the United States once Germany was its enemy. The scientific benefit of war, particularly of confiscated technologies from enemies, however, has had very little written about it.

Electron. There are several books on the discovery of the electron, but most are academic in nature, such as *A History of the Electron: J. J. and G. P. Thomson* (Jaume Navarro), *Flash of the Cathode Ray: A History of J. J. Thomson's Electron* (Per F. Dahl), *Electron: A Centenary Volume* (Michael Springford), and *J. J. Thomson and the Discovery of the Electron* (E. A. Davis and Isobel Falconer). These titles are not intended for general audiences, and very few give narrative accounts of the discovery, but readers will be able to glean the impact of J. J. Thomson's work from their pages nevertheless. The best sources of materials for general readers are in magazine articles and short biographical profiles within the scientific literature. The article "Sir J. J. Thomson, O. M., FRS: A Centenary Biography" written by D. J. Price for *Nuovo Cimento* (1956), as well as "J. J. Thomson and the Discovery of the Electron" written by George Paget Thomson in *Physics Today* (1956) are accounts that are closer to describing the impact of this man's work to nonscientists. J. J. Thomson's son, George Paget Thomson (who was an esteemed scientist in his own right), dutifully kept the memory of his father alive with various articles. Many of them provide the same depictions, however. Interestingly, one of them, called "J. J. Thomson as We Remember Him," was cowritten with George's sister Joan and provides fresh insight into J. J.'s personality.

J. J. Thomson wrote an autobiography called *Recollections and Reflections*. Unfortunately, J. J. never kept a diary, so his childhood is still somewhat of a mystery. Nevertheless, he did a solid job of providing insights into his upbringing and his discoveries. (J. J. Thomson had strong opinions on how to teach science, which are illuminating to read.) There are a few older biographies on J. J. Thomson, one of which is *The Life of Sir J. J. Thomson: Sometime Master of Trinity College, Cambridge* by Lord Rayleigh. This book does a thorough job of describing the man's life and his work, and is probably the best resource for such information. For a contemporary discussion of J. J.'s work, Isobel Falconer's dissertation on Thomson, as well as her article and book *J. J. Thomson and the Discovery of the Electron*, are worth acquiring. Lastly, a good summary that gives context of the world of physics at the time of J. J. Thomson can be found in the introduction to Emilio Segrè's *From X-rays to Quarks: Modern Physicists and Their Discoveries* published by W. H. Freeman and Co.

While there is much written about J. J. Thomson, there is very little on Ebeneezer Everett, unfortunately. To offset this, J. J. Thomson attempted to mark this man's contribution to science when he wrote Everett's obituary in one of the most highly regarded science magazines in the UK, *Nature*. From this passage, J. J.'s respect for

Everett is crystal clear. The importance of technicians to scientists was often an untold secret in the world of science, but this knowledge is finally coming to light. One scientific publication on this topic is "Keeping the Culture Alive: The Laboratory Technician in Mid-Twentieth Century British Medical Research" by E. M. Tansey.

Chapter 8: Think

Phineas Gage. Phineas Gage is a medical patient discussed in many introductory psychology and neuroscience textbooks. Even 150 years after Gage's accident, a recent report has appeared in *Science,* written by Hanna Damasio and coworkers. These researchers used modern medical tools on Gage's skull to ascertain specifically where he was injured, since no autopsy was done on him at the time of his death. This recent paper "The Return of Phineas Gage" will get any reader up to speed on the current state of the medical world's understanding of Gage's prognosis. For a curious reader, though, the original medical articles by Dr. John Harlow (1848, 1849, and 1868) and Dr. Henry Bigelow (1850), along with newspaper articles from Vermont, provide the closest accounts at the time of the accident. For readers wanting to know all there is about Gage, the most exhaustive book to date is *An Odd Kind of Fame* by Malcolm Macmillan. Macmillan's book includes some of the key medical papers mentioned above in its appendix; there is also plenty of original research. This book is not a breezy read, however. The author chronicles events in a descriptive, nonnarrative form, which is possibly due to that paucity of archival materials or collected papers on Gage. Nevertheless, Macmillan's book is wonderfully useful for those wanting to know more about neuroscience's patient zero.

George Willard Coy. Given the importance of George W. Coy's telephone exchange, the amount of information about Coy and his invention is scarce. The origin story can be found in the pages of rare and old books, such as *Connecticut Pioneers in Telephony* by John Leigh Walsh and *The First Century of the Telephone in Connecticut* by Reuel A. Benson, Jr., as well as the *Popular Science Monthly* article from January 1907 entitled "Notes on the Development of Telephone Service III." Most of these sources are housed in the New Haven Museum, the Connecticut Historical Society, and the Connecticut State Library. The workings of the switchboard are best described in the book *Race on the Line: Gender, Labor, and Technology in the Bell System* by Venus Green on pages 20 and 21. For those who desire the electrical schematics of the switchboard, the appendix to Walsh's *Connecticut Pioneers in Telephony* will certainly satisfy this need. The best place to acquire information about the first telephone exchange is the New Haven Museum (which has a vertical file as well as a replica of the switchboard in its exhibits). Since the state of Connecticut was the first to establish the telephone business, various local newspaper articles have been written on key anniversaries. Interestingly, the Boardman Building, where Coy's exchange began, was designated a historical landmark until 1973, when it was demolished; train tracks now occupy the site, and there is no historical plaque for it. Coy is slowly getting the credit he deserved, however. In 2017, the theater company Broken Umbrella produced a play about him called *Exchange*. Despite such efforts, Coy still remains a little-known part of Connecticut and American history.

Almon Strowger. Almon Strowger is telephone's forgotten man, so the literature on him is limited. For information about his invention, there is a short entry in *Inventing the 19th Century* by Stephen van Dulken, which describes Strowger's work. For an account about Strowger's life as well as his invention, the book *Good Connections: A Century of Service by the Men & Women of Southwestern Bell* by David G. Park, Jr., has some things to offer as well. (Both of these items can be acquired from the Kansas City Public Library's vertical files.) Also, Lewis Coe's *Telephone and Its Several Inventors: A History* is well worth finding for its entries on Strowger and many other inventors. For more materials than what is found in books, a researcher would do well to contact the La Porte Historical Society as well as the Penfield Historical Society for their materials. Lastly, numerous newspaper articles written between 1899 and 1902 about Strowger pertain to saying goodbye to the "hello girls."

The Birth of the Transistor. A number of books discuss the birth of the transistor. The most seminal work, *Crystal Fire: The Invention of the Transistor and the Birth of the Information Age* by Michael Riordan and Lillian Hoddeson, is a well-researched and well-written book that sets the tone for how to tell this story superbly. More recent books on the topic of semiconductors include *The Chip: How Two Americans Invented the Microchip and Launched a Revolution* by T. R. Reid, *The Innovators: How a Group of Hackers, Geniuses, and Geeks Created the Digital Revolution* by Walter Issacson, and *The Idea Factory: Bell Labs and the Great Age of American Innovation* by John Gertner, each adding specifics and continuing the tradition of great exposition. For a more technical account of the birth of our silicon age, the book *The Electronic Genie: The Tangled History of Silicon* is cowritten by Frederick Seitz, one of the scientists who created this modern wonder. There is also *Sand and Silicon: Science that Changed the World* by Denis McWhan, which shares with readers anything they could ever want to know about this element and the way it serves society.

Information about the physics of semiconductors can be readily found in a number of materials-science textbooks; however, they may be overly technical for most readers. Fortunately, readable texts about the structure of crystals and their properties were produced decades ago by Bell Laboratories. One key author, Alan Holden, had a gift for making complex concepts clear and accessible to lay readers. Titles worth acquiring are *The Nature of Solids* and *Conductors and Semiconductors,* both by Holden. *Half-Way Elements* by Graham Chedd is a hard-to-find paperback, but written in a popular and clear style. In addition to these older efforts to make semiconductors understandable, there has been a recent thrust with a similar mission. The text *The Substance of Civilization* by Stephen L. Sass falls into this category, as well as the valiant effort made by Rolf E. Hummel in his *Understanding Materials Science: History, Properties, Applications* textbook. There is no cartoon guide to materials science, but there should be. Until then, a great, but dated, movie called *Silicon Run* shows the makings of modern integrated circuits, which helps the viewer appreciate all the steps involved in making the heart of our cell phones and computers.

Impact of the Internet. While the impact of the internet is still new, some early scientific papers point to how the brain is being changed by this invention. The 2011 research paper in *Science* entitled "Google Effects on Memory: Cognitive Consequences of Having Information at Our Fingertips," written by Harvard researchers Betsy Sparrow and her coworkers, was an early scientific clarion call to the effects of our devices on us. This work, while important, may not have reached the general public as it should. Fortunately, an article written for *The Atlantic* by Nicholas Carr dropped a bombshell: "Is Google Making Us Stupid?" Carr later wrote a book called *The Shallows: What the Internet Is Doing to Our Brains* that expounded on this theme. Using a mix of first-person experiences along with science narrative, he created an erudite yet approachable tome. The book was a finalist for the Pulitzer Prize and the exposition and research make it clear why.

Other books provide some foundational materials for how our brains are being changed by the internet. One book in particular by Torkel Klingberg called *The Overflowing Brain: Information Overload and the Limits of Working Memory* takes a step-by-step approach to discussing how the brain works in the process of storing information, how working memory—the scratch paper of our brains—has a limit, and how that limit has been reached with our time on the web. Another book that discusses the impact of computers on our brains is Nicholas Kardaras's *Glow Kids: How Screen Addiction Is Hijacking Our Kids—and How to Break the Trance*. There is also James Gleick's award-winning book *The Information*, which illustrates in great detail how the deluge of information has shaped humans.

A spate of efforts shows how the internet changes different aspects of society. Charles Seife's *Virtual Unreality: The New Era of Digital Deception* talks about the unreliability of information on the internet, predicting in some ways the problem of fake news on the web; also, Michael Patrick Lynch's *The Internet of Us: Knowing More and Understanding Less in the Age of Big Data* makes a case for the difference between *knowing* and *Google-knowing*. Scott Timberg's *Culture Crash: The Killing of the Creative Class* looks at the role of art-ists in the information age. Decades earlier, in 1995, Clifford Stoll's *Silicon Snake Oil: Second Thoughts on the Information Highway* showed how the internet was changing us. In this book, Stoll shared his second thoughts on the World Wide Web just a few years before Google was born.

Cognitive scientists are finding clever ways for our devices to keep our attention. The technical book *Human Attention in Digital Environments*, edited by Claudia Roda and published by Cambridge University Press, is beyond the purview of the everyday reader, but if scanned, one can see that human attention is being corralled and controlled with computer interactions. Cognitive scientists are learning more about how we think and how to manage how we think when interacting with computers. This fact alone should give a reader pause.

Many books and articles speak to the brain and creativity. The work on how the internet affects creativity is still very new, how-ever. Nevertheless, some key works elucidate how creativity happens and how it might be influenced by the internet. Kenneth Heilman's article "Possible Brain Mechanism of Creativity" discusses the parts of the brain that are active for different creative endeavors. His book *Creativity and the Brain* is sweeping in scope, but doesn't get to the topic of creativity until the last chapter. For an introduction to the topic of creativity and the brain, there are also Wlodzislaw Duch's article "Creativity and the Brain" and Nancy Coover Andreasen's book *The Creating Brain: The Neuroscience of Genius*. Andreasen wrote an excellent primer on the topic, from brain plasticity to mental exercises for becoming more creative. It should be noted that the topic of the interactions of creativity, the brain, and the internet is still very new, so there is still much to learn and understand. What all researchers agree upon is that creativity requires flow, a topic written about expertly by Mihaly Csikszentmihalyi in his book entitled *Flow: The Psychology of Optimal Experience*.

Technology and Humans. Every decade a few books pop up in the literature that examine society and technology; some look at technology with wonder and others with worry. In the twentieth century, Alex Broers's monograph *The Triumph of Technology* looks fondly at technology. A century earlier, Hendrik van Loon's *The Story of Inventions: Man, the Miracle Maker* showed how tools made by our earliest ancestors enabled humankind to do so much. In many ways, these books are right to look at invention from that vantage point. In our advanced times, however, we know that technology need not be viewed just as something gained and something lost. More recent books take a Schrödinger's cat approach, with both antithetical states coexisting. One such book that takes a balanced approach of technophilia and technophobia is *Living with the Genie: Essays on Technology and the Quest for Human Mastery* edited by Alan Lightman, Daniel Sarewitz, and Christina Desser.

For books that are much more pessimistic about technology's impact on our present and our future, there is the erudite examination of the changes to society in *The Technological Society* by Jacques

Ellul. Lewis Mumford's *Technics and Civilization* takes a matter-of-fact perspective toward how technology shaped us, as does Marshall McLuhan's *Understanding Media: The Extension of Man*. This latter book is certainly a "must read," for McLuhan is prophetic at times; but this book is not always a "must understand," for McLuhan prided himself on clever—but not necessarily understandable—prose. Books written by futurist Alvin Toffler, such as *Future Shock* and *Third Wave*, have resonated with readers by illuminating the feeling of "too much change at once," and by giving a name to the "information overload" most were experiencing. Such books are outdated in some parts, but up to date in others.

All in all, *The Alchemy of Us* is of the ilk of books that serve society with a call to action. The most recent of these offerings is *The Shallows* by Nicholas Carr, which is the descendant of Rachel Carson's *Silent Spring,* the seminal book that set the tone for how to examine our creations a generation ago. As *The Alchemy of Us* shows, we can certainly love technology, but we must not be enchanted by it. Real love accepts faults, but also seeks to correct them. This is the mindset that lies at the heart of this book, and the mission for its writing. Technology and humankind must cocreate, but not at the cost of humanity itself.

Quote Permissions

Illustration Credits

1: Fox Photos/Getty Images

2: The Worshipful Company of Clockmakers' Collection, UK/ Bridgeman Images

3–5: Courtesy Sheffield Archives and Local Studies www.pictureshef-field.com

6–11, 89–92, 94, 95, 97, 98, 100: Courtesy of AT&T Archives and History Center

12, 14–16, 20, 29–32, 34, 40, 42, 44–45, 60, 96: Library of Congress

13: Chicago History Museum, ICHi-176199

17: Angela Pitaro's collection

18–19: Illustrated by Mark Saba after Atlas of Historical Geography of the United States, used with permission

21: National Portrait Gallery, London

22: National Portrait Gallery, Smithsonian Institution

23: Artist: Mark Saba.

24, 88, 93: Public Domain: Wikipedia; New Haven Free Public Library: CT State Libraries

25, 43, 49, 61, 67, 74, 80: Author's collection

26: Courtesy of the Smithsonian Libraries, Washington, DC

27: Scientists and Inventors Portrait File, Archives Center, National Museum of American History, Smithsonian Institution

28, 50, 101: Division of Work and Industry, National Museum of American History, Smithsonian Institution

33: Special Collections, University of Virginia, Charlottesville, VA

35–36, 38–39: Courtesy of the Department of Special Collections, Stanford Libraries

37: By permission of Kingston Museum and Heritage Services

41: Courtesy of the George Eastman Museum

46: Copyright Guardian News & Media Ltd. 2018

47: Courtesy of POLOMAD

48, 51: Courtesy of The Derby Historical Society

52–53, 56, 59: U.S. Department of the Interior, National Park Service, Thomas Edison Historical Park

54: Chicago History Museum, ICHi-176200; Photography by David H. Anderson

55: The Thomas A. Edison Papers at Rutgers University

57–58: Courtesy NASA/JPL-Caltech, with permission from John Casani

62, 102: Gordon Library Archives and Special Collections at the Worcester Polytechnic Institute (all three images)

63–65, 68, 69, 71–73: Courtesy of International Business Machines Corporation, Copyright International Business Machines Corporation

66: Courtesy of the Science History Institute

70: Gordon Library Archives and Special Collections at the Worcester Polytechnic Institute

75–77: Alexander Fleming Laboratory Museum (Imperial College Healthcare NHS Trust)

Index

About the Author

Ainissa Ramirez, PhD, is an award-winning scientist and science communicator. A graduate of Brown University, she earned her doctorate in materials science and engineering from Stanford. Dr. Ramirez started her career as a scientist at Bell Laboratories in Murray Hill, New Jersey, and later worked as an associate professor of mechanical engineering at Yale. She authored the books *Save Our Science* and *Newton's Football*. She has written for *Forbes*, *Time,* and *Science* and has explained science headlines on CBS, CNN, NPR, and PBS. She speaks widely on the topics of science and technology and gave a TED talk on the importance of science education.

Ainissa Ramirez lives in New Haven, Connecticut (www.ainissaramirez.com).